DISCARD

GPRS
(General Packet Radio Service)

MCGRAW-HILL
TELECOMMUNICATIONS

GPRS

Regis J. (Bud) Bates

McGraw-Hill
New York   Chicago   San Francisco   Lisbon
London   Madrid   Mexico City   Milan   New Delhi
San Juan   Seoul   Singapore   Sydney   Toronto

McGraw-Hill

A Division of The McGraw·Hill Companies

1 2 3 4 5 6 7 8 9 0   AGM/AGM   9 8 7 6 5 4 3 2 1

ISBN 0-07-138188-0

The sponsoring editor for this book was Steve Chapman and the production supervisor was Pamela Pelton. It was set in Century Schoolbook by MacAllister Publishing Services, LLC.

Printed and bound by Quebecor/Martinsburg.

Throughout this book, trademarked names are used. Rather than put a trademark symbol after every occurrence of a trademarked name, we use names in an editorial fashion only, and to the benefit of the trademark owner, with no intention of infringement of the trademark. Where such designations appear in this book, they have been printed with initial caps.

 This book is printed on recycled, acid-free paper containing a minimum of 50 percent recycled de-inked fiber.

ABOUT THE AUTHOR

Regis J. (Bud) Bates Jr., President
TC International Consulting, Inc.
Phoenix, Arizona
1-800-322-2202

Regis (Bud) Bates has more than 35 years of experience in telecommunications and management information systems (MIS). He oversees the operation of TC International Consulting, Inc., a full service management consulting organization. He has been involved in the design of major networks including LANs and WANs. His clients span the range of Fortune 100–500 companies. Many of his projects deal with multiple sites and countries using Frame Relay, ATM, and Optical architectures. He has also done a significant amount of work in the wireless communications area.

Bud also develops and conducts various public and in-house seminars ranging from a managerial overview to very technical instruction on voice, data, LAN, WAN, and broadband communications. For the past two years, he has devoted much of his development and training activities on the convergence of voice and data communications. Included in these developments, Bud has been training numerous CLECs on the integration of voice and data. He has recommended and implemented several training programs (in-house) using all the technologies that are converging as a base model. Included in this list are several training programs that carry the organization's internal certification. His many topics include both basic and advanced courseware on voice, data, LAN, WAN, ATM, SONET, T1/T3, VoIP, and Voice over Data Protocols (FR, ATM, and so on).

Bud has written numerous books on the technologies, many of which have been best sellers for McGraw-Hill. Moreover, his *Voice and Data Communications Handbook* has led McGraw-Hill's sales for three consecutive years, with a new revision released in August 2001. His recent publication *Broadband Telecommunications Handbook* (December 1999) has been an equal seller. Some of his other titles include, *Introduction to T1/T3 Networking*; *Disaster Recovery for LANs: A Planning and Action Guide*; *Telecommunications Disaster Recover*; *Wireless Networked Communications: Concepts, Systems, and Implementation*; *Optical Switching and Networking*; *Nortel Networks Layer 3 Switching*; and *Wireless Broadband Handbook*.

Bud also works with venture capitalists for various analyses and studies. One of his recommendations got the investors to increase the recommended funding from $100 million to more than $400 million. He has consistently been on the mark with his projections.

CONTENTS

Contents

Contents

Contents

ACKNOWLEDGMENTS

Well, here I am again, finalizing the latest book for McGraw-Hill. This one is different because it focuses on a single subject! The idea for this book came earlier this year, but has blossomed quite a bit since then because of the progress being made in the industry.

This book deals with the issues surrounding the data and wireless industry as the two converge. I hope that this material is comfortable for you, the reader, because there is so little published about GPRS for the novice and business professional.

To be fair about this production, I owe a lot of credit to many people. Some of the people interacted with me regularly, others only occasionally. First, I must thank Steve Chapman, who is McGraw-Hill's executive editor on this book and a super person. Steve knew that the deadlines were approaching, but kept his cool and kept after me. Numerous other people aided in the editing and production of the book—far too many to name. They know who they are and can give themselves a pat on the back for their efforts. Also worth special mention is Molly Applegate who has worked with me on two other books. Keeping track of books from the same author has to be a challenge in itself, especially when many of the topics seem to run together. Molly's usual tenacity got the final production together. Then there are the vendors and manufacturers we talk to daily regarding products, services, and opportunities. All add to the knowledge in this book.

Two people in my office deserve the lion's share of the credits for the ultimate graphical representation and production of this book. First is Gabriele, who is my wife and partner of more than three decades. She is the steady contributor to my production. Gabriele truly deserves to have her name on the cover of this book because she has been a partner in the production of the preceding 14 books and barely gets mention. Second is a young lady whose energy and enthusiasm I have mentioned before: Amber Hartmann (our specialty graphics person). Amber and Gabriele worked dauntlessly to keep the graphics accurate and readable while at the same time creating interesting renderings. All this, despite the technical aspects of the drawings, had to be maintained. On more than one occasion, these two ladies were cursing the technical drawings because of their complexity. Yet, they both came through in record time.

This team of people all pulled together to make *GPRS* a reality. They all deserve the credit more than I.

As usual, I want to especially thank you, the reader, for giving up your time to read this book. I receive many calls and e-mails from readers who just want to let me know that they enjoyed my opinion or the way I presented an idea. I hope I can continue to win your support. My best wishes to you all!

Introduction
to GSM

Objectives

When you complete the reading in this chapter, you will be able to

- Describe the main components of a GSM network.
- Describe the mobile services.
- Understand how a mobile performs an attach or detach procedure in GSM.
- Discuss the modulation techniques used for GSM.
- Understand the access methods used.
- Describe the overall cellular operation of a radio network.

Welcome to an overview of the *General Packet Radio Services* (GPRS). GPRS is a radio service that was designed to run on *Global Systems for Mobile* (GSM), a worldwide standard for cellular communications. Data transmissions in the past were slow across the radio interfaces due to many propagation and reception problems. To create a broadband communications interface, GPRS was developed as a stepping-stone approach to other services like the *Enhanced Data for a Global Environment* (EDGE). Regardless of the names we place on these services, the real issues are how much (cost) and how fast (speed) we need to meet the demands for data transmission now and in the future.

Before delving directly into the GPRS systems and services, it is prudent to have common ground on the use of the radio-based systems. Therefore, a review (or introduction) of GSM is appropriate. After all, if GPRS is an overlay to GSM, we should at least understand how and why GSM works.

History of Cellular Mobile Radio and GSM

The idea of cell-based mobile radio systems appeared at Bell Laboratories in the early 1970s. However, the commercial introduction of cellular systems did not occur until the 1980s. Because of the pent-up demand and newness, analog cellular telephone systems grew rapidly in Europe and North America. Today, cellular systems still represent one of the fastest growing telecommunications services. Recent studies indicate that three of four new phones are mobile phones. Unfortunately, when cellular systems were first being deployed, each country developed its own system, which was problematic because

- The equipment only worked within the boundaries of each country.
- The market for mobile equipment manufacturers was limited by the operating system.

Three different services had emerged in the world at the time. They were

- *Advanced Mobile Phone Services* (AMPS) in North America
- *Total Access Communications System* (TACS) in the United Kingdom
- *Nordic Mobile Telephone* (NMT) in Nordic countries

To solve this problem, in 1982 the Conference of *European Posts and Telecommunications* (CEPT) formed the *Groupe Spécial Mobile* (GSM) to develop a pan-European mobile cellular radio system (the acronym later became Global System for Mobile communications). The goal of the GSM study group was to standardize systems to provide

- Improved spectrum efficiency
- International roaming
- Low-cost mobile sets and base stations
- High-quality speech
- Compatibility with *Integrated Services Digital Network* (ISDN) and other telephone company services
- Support for new services

The existing cellular systems were developed on analog technology. However, GSM was developed using digital technology.

Benchmarks in GSM

Table 1-1 shows many of the important events in the rollout of the GSM system; other events were introduced, but had less significant impact on the overall systems.

Commercial service was introduced in mid-1991. By 1993, 36 GSM networks were already operating in 22 countries. Today, you can be instantly reached on your mobile phone in over 171 countries worldwide and on 400 networks (operators). Over 550 million people were subscribers to GSM

Table 1-1

Major Events in
GSM

Year	Events
1982	CEPT establishes a GSM group in order to develop the standards for a pan-European cellular mobile system.
1985	A list of recommendations to be generated by the group is accepted.
1986	Field tests are performed to test the different radio techniques proposed for the air interface.
1987	*Time Division Multiple Access* (TDMA) is chosen as access method (with *Frequency Division Multiple Access* [FDMA]). The initial *Memorandum of Understanding* (MoU) is signed by telecommunication operators representing 12 countries.
1988	GSM system is validated.
1989	The responsibility of the GSM specifications is passed to the *European Telecommunications Standards Institute* (ETSI).
1990	Phase 1 of the GSM specifications is delivered.
1991	Commercial launch of the GSM service occurs.
1992	The addition of the countries that signed the GSM Memorandum of Understanding takes place. Coverage spreads to larger cities and airports.
1993	Coverage of main roads GSM services starts outside Europe.
1995	Phase II of the GSM specifications occurs. Coverage is extended to rural areas.

mobile telecommunications.[1] GSM truly stands for Global System for Mobile telecommunications. Roaming is the ability to use your GSM phone number in another GSM network. You can roam to another region or country and use the services of any network operator in that region that has a roaming agreement with the GSM network operator in your home region/country. A roaming agreement is a business agreement between two network operators to transfer items such as call charges and subscription information back and forth as their subscribers roam into each other's areas.

---

[1] As of May 2001

GSM Metrics

The GSM standard is the most widely accepted standard and is implemented globally, owning a market share of 69 percent of the world's digital cellular subscribers. TDMA, with a market share close to 10 percent, is available mainly in North America and South America. GSM, which uses a TDMA access, and North American TDMA are two of the world's leading digital network standards. Unfortunately, it is currently technically impossible for users of either standard to make or receive calls in areas where only the other standard is available. Once interoperability is in place, users of GSM and TDMA handsets will be able to roam on the other network type —subject to the agreements between mobile operators. This will make roaming possible across much of the world because GSM and TDMA networks cover large sections of the global population and together account for 79 percent of all mobile subscribers, as shown in Figure 1-1.

Cell Structure

In a cellular system, the coverage area of an operator is divided into cells. A cell is the area that one transmitter or a small collection of transmitters can cover. The size of a cell is determined by the transmitter's power. The concept of cellular systems is the use of low-power transmitters in order to enable the efficient reuse of the frequencies. The maximum size of a cell is approximately 35 km (radius), providing a round-trip communications path

Figure 1-1
Market penetrations
of GSM and TDMA.

As of May, 2001

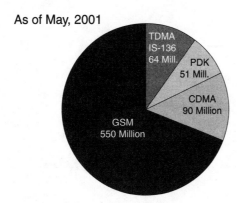

from the mobile to the cell site and back. If the transmitters are very powerful, the frequencies cannot be reused for hundreds of kilometers, as they are limited to the coverage area of the transmitter. In the past when a mobile communications system was installed, the coverage blocked the reuse beyond the 25-mile coverage area, and created a corridor of interference of an additional 75 miles. This is shown in Figure 1-2.

The frequency band allocated to a cellular mobile radio system is distributed over a group of cells and this distribution is repeated in all of an operator's coverage area. The entire number of radio channels available can then be used in each group of cells that form the operator's coverage area. Frequencies used in a cell will be reused several cells away. The distance between the cells using the same frequency must be sufficient to avoid interference. The frequency reuse will increase the capacity in the number of users considerably. The patterns can be a four-cell pattern or other choices. The typical clusters contain 4, 7, 12, or 21 cells.

In order to work properly, a cellular system must verify the following two main conditions:

■ The power level of a transmitter within a single cell must be limited in order to reduce the interference with the transmitters of neighboring

Figure 1-2
The older way of handling mobile communications.

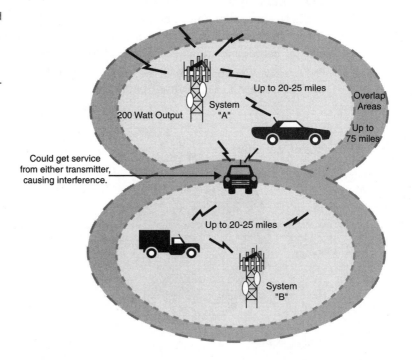

cells. The interference will not produce any damage to the system if a distance of about 2.5 to 3 times the diameter of a cell is reserved between transmitters. The receiver filters must also conform.

■ Neighboring cells cannot share the same channels. In order to reduce the interference, the frequencies must be reused only within a certain pattern. The pattern may also be a seven-cell pattern, which is shown in Figure 1-3.

In order to exchange the information needed to maintain the communication links within the cellular network, several radio channels are reserved for the signaling information. Sometimes we use a 12-cell pattern with a repeating sequence. The 12-cell pattern is really a grouping of three four-cell clusters, as shown in Figure 1-4. The larger the cell pattern, the more the coverage areas tend to work. In general, the larger cell patterns

Figure 1-3
The seven-cell
pattern.
(Source: ETSI)

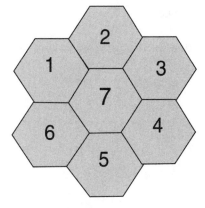

Figure 1-4
The 12-cell pattern.
(Source: ETSI)

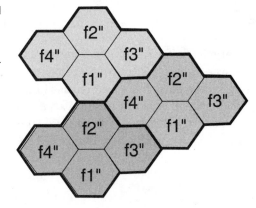

are used in various reuse patterns to get the most out of the scarce radio resources as possible. The 21-cell pattern is by far the largest repeating pattern in use today. The cells are grouped into clusters. The number of cells in a cluster determines whether the cluster can be repeated continuously within the coverage area.

The number of cells in each cluster is very important. The smaller the number of cells per cluster, the greater the number of channels per cell. Therefore, the capacity of each cell will be increased. However, a balance must be found in order to avoid the interference that could occur between neighboring clusters. This interference is produced by the small size of the clusters (the size of the cluster is defined by the number of cells per cluster). The total number of channels per cell depends on the number of available channels and the type of cluster used.

Types of Cells

The density of population in a country is so varied that different types of cells are used:

- Macrocells
- Microcells
- Selective or sectorized cells
- Umbrella cells
- Nanocells
- Picocells

Macrocells

Macrocells are large cells for remote and sparsely populated areas. These cells can be as large as 3 to 35 km from the center to the edge of the cell (radius). The larger cells place more frequencies in the core, but because the area is rural, the macrocell typically has limited frequencies (channels) and higher-power transmitters. This is a limitation that prevents other sites from being closely adjacent to this cell. Figure 1-5 shows the macrocell.

Figure 1-5
The macrocell.

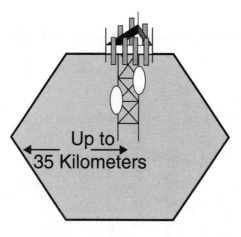

Microcells

These cells are used for densely populated areas. By splitting the existing areas into smaller cells, the number of channels available and the capacity of the cells are increased. The power level of the transmitters used in these cells is then decreased, reducing the possibility of interference between neighboring cells. Some of the microcells may be as small as .1 to 1 km depending on the need. Often times the cell splitting will use the reduced power and the greater coverage to satisfy hot spots or dead spots in the network.

Another need may well be a below-the-rooftop cell that satisfies a very close-knit group of people or varied users. The picocell will be in a building, and is typically a smaller version of a microcell. The distances covered with a picocell are approximately .01 to 1 km. These are used in office buildings for close in calling, part of a *Private Branch Exchange* (PBX) or a wireless *Local Area Network* (LAN) application today. A small group of users will share this cell because of the close proximity to each other and larger cells around. Nanocells also fall into the below-the-rooftop domain where the distances for this type of cell are from .01 to .001 km. These are just smaller and smaller segments that are built within a building, as an example. Figure 1-6 shows a combination of a microcell and picocell.

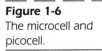

Figure 1-6
The microcell and
picocell.

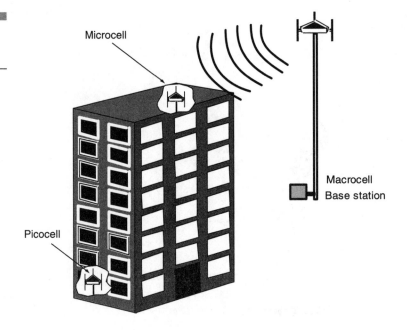

Selective Cells or Sectorized Cells

It is not always useful to define a cell with a full coverage of 360 degrees. In some cases, cells with a particular shape and coverage are needed. These cells are called selective cells. Selective cells are typically the cells that may be located at the entrances of tunnels where 360 degrees coverage is not needed. In this case, a selective cell with coverage of 120 degrees is used. This selective cell is shown in Figure 1-7.

Tiered Cells

A tiered cell is one where an overlay of radio equipment operates in two different frequencies and uses different sectors. The tiered cell is also a form of a selective cell.

Umbrella Cells

Alongside of a high-speed freeway, crossing very small cells produces an overabundance of handovers among the different small neighboring cells.

Figure 1-7
The selective cell.

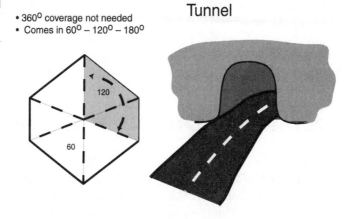

- 360° coverage not needed
- Comes in 60° – 120° – 180°

Tunnel

To solve this problem, the concept of umbrella cells was introduced. An umbrella cell covers several microcells, as shown in Figure 1-8. The power level inside an umbrella cell is increased compared to the power levels used in the microcells that form the umbrella cell. How does the cell know when to shift from a microcell to an umbrella cell? When the speed of the mobile is too high, the mobile is handed off to the umbrella cell. The mobile will then stay longer in the same cell (in this case, the umbrella cell). This will reduce the number of handovers and the work of the network. A large number of handover demands and the propagation characteristics of a

Figure 1-8
The umbrella cell.

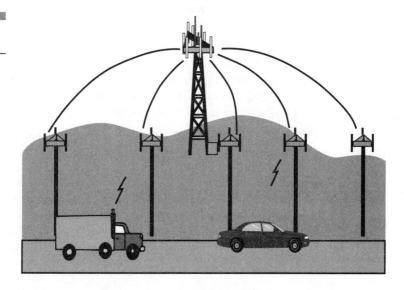

mobile can help to detect its high speed. The radio equipment is no longer forced to constantly change hands from cell to cell when using this umbrella. This meets the goal of GSM in that the efficient use of the *radio frequency* (RF) spectrum is what is being achieved.

Analog to Digital Movement

In the 1980s, most mobile cellular systems were based on analog systems including AMPS, TACS, and NMT. In fact, 95 percent of the United States has coverage from AMPS services, whereas only 70 percent is covered with digital service. The roaming agreements used between cellular carriers in North America use AMPS for roaming. In many cases, the analog networks are starting to wind down in the major metropolitan areas; however, in rural communities, AMPS is still predominant. GSM system was the first digital cellular system created from the onset. Different reasons explain the transition from analog to digital technology. Cellular systems experienced phenomenal growth. Analog systems were not able to cope with this increasing demand. To overcome this problem, new frequency bands and new technologies were suggested. Many countries rejected the possibility of using new frequency bands because of the restricted spectrum (even though later on, other frequency bands were allocated for the development of mobile cellular radio). New analog technologies were able to overcome some of the problems, but were too expensive. The digital radio was the best option (but not the perfect one) to handle the capacity needs in a cost-efficient manner.

The decision to adopt digital technology for GSM was made in the course of developing the standard. During the development of GSM, the telecommunications industry converted to digital networking standards. ISDN is an example of this evolution. In order to make GSM compatible with the services offered by ISDN, it was decided that digital radio technology was the best option available.

Quality of service can also be improved dramatically by using digital rather than analog technology. From the beginning, the planners of GSM wanted ISDN compatibility in the services offered and control signaling used. The radio link imposed some limitations because the standard ISDN bit rate of 64 Kbps could not be practically achieved.

Using the *International Telecommunication Union-Telecommunication Standardization* (ITU-T) definitions, telecommunication services can be divided into the following categories:

- ■ Teleservices
- ■ Bearer services
- ■ Supplementary services

Teleservices

The most basic teleservice supported by GSM is telephony, the transmission of speech. It has an added emergency service, where the nearest emergency service provider is notified by dialing three digits. The emergency number 112 is used like 911 in North America. Group 3 fax, an analog method described in ITU-T recommendation T.30, is also supported with the use of an appropriate fax adapter.

A unique feature of GSM compared to older analog systems is the *Short Message Service* (SMS). SMS is a bidirectional service for sending short alphanumeric (up to 160 bytes) messages in a store-and-forward fashion. For point-to-point SMS, a message can be sent to another subscriber to the service, and an acknowledgement of receipt is provided to the sender. SMS can also be used in a cell broadcast mode for sending messages such as traffic updates or news updates. Messages can be stored in the *Subscriber Identity Module* (SIM) card for later retrieval. The SMS service has been very well accepted with over 1 billion SMS messages being sent monthly.

As things progressed, Phase II of GSM introduced enhancements. For example, in the teleservices, half-rate voice coding was introduced. In the first phase, full-rate voice coding was used at a rate of 13 Kbps for a voice conversation. Later, the 6.5-Kbps vocoders were introduced for use at a network operator's choice. This enables the network operator to offer good speech quality to twice as many users without any additional radio resources. Essentially, we split the channel in half, because people actually carry traffic on the channel only 25 to 30 percent of the time.

Enhancements also included better SMS informational flow for point-to-point communications and the use of point-to-multipoint communications. The 160-character SMS message was finally documented and became fully store-and-forward.

Bearer Services

The digital nature of GSM enables data, both synchronous and asynchronous, to be transported as a bearer service to or from an ISDN terminal.

Data can use either the transparent service, with a fixed delay but no guarantee of data integrity, or a nontransparent service, which guarantees data integrity through an *Automatic Repeat Request* (ARQ) mechanism, which unfortunately introduces a variable delay. The data rates supported by GSM are 300 bps, 600 bps, 1,200 bps, 2,400 bps, and 9,600 bps. One can imagine in this new millennium that these data speeds are intolerable for the mainstay of data transmission. In fact, if someone were to offer us Internet access at speeds of up to 9,600 bps, we would probably become very disinterested in the service. Yet, from a mobile perspective, these speeds were considered quite fast.

Enhancements from Phase II also included better throughput for data transmission of data using a synchronous dedicated packet data access operating at 2.4 to 9.6 Kbps. Phase I only accepted an asynchronous access to a dedicated *packet assembler/disassembler* (PAD). The access of data through a dedicated PAD at the entrance of an X.25 network enables access to a higher degree of reliable data transport, helping to overcome the link layer problems on the radio.

Data is now available over the GSM Phase II at both send and receive speeds of up to 9.6 Kbps. In the earlier releases, slower data was more prevalent. The use of the GSM network enables the integration of various network platforms such as

- *Plain Old Telephone Services* (POTS)
- ISDN access and emulation
- Packet data network access (X.25 and IP are the most common)
- Circuit-switched data transfer across and X.25, X.31, and X.32 standard

Because the data is being sent across a digital air interface, no modem is required at the *mobile station* (MS) end.

Supplementary Services

Supplementary services (which are really the added features of the cellular networks) are provided on top of teleservices or bearer services, and include features such as

- Caller identification.
- Call forwarding; the subscriber can forward incoming calls to another number if the called mobile is busy (CFB), unreachable (CFNRc), or if

no reply (CFNRy) occurs. Call forwarding can also be applied unconditionally (CFU).

■ Call waiting.

■ Multiparty conversations.

■ Barring of outgoing (international) calls. Different types of call barring services are available:

- *Barring of All Outgoing Calls* (BAOC)

- *Barring of Outgoing International Calls* (BOIC)

- *Barring of Outgoing International Calls except those directed toward the Home PLMN Country* (BOIC-exHC)

- *Barring of All Incoming Calls* (BAIC)

- *Barring of Incoming Calls when Roaming* (BAIC-R)

Phase II enhancements to the supplementary services include the following:

■ *Calling/Connected Line Identification Presentation* **(CLIP)**
This supplies the called user with the ISDN of the calling user.

■ *Calling/Connected Line Identification Restriction* **(CLIR)** This enables the calling user to restrict the presentation.

■ *Connected Line identification Presentation* **(CoLP)** This supplies the calling user with the directory number he or she receives if his or her call is forwarded.

■ *Connected Line identification Restriction* **(CoLR)** This enables the called user to restrict the presentation.

■ *Call Waiting* **(CW)** This informs the user, during a conversation, about another incoming call. The user can answer, reject, or ignore this incoming call.

These are added supplementary services finishing off the list:

■ **Call hold** This puts an active call on hold.

■ *Advice of Charge* **(AoC)** This provides the user with online charge information.

■ **Multiparty service** This creates the possibility of establishing a multiparty conversation.

■ *Closed User Group* **(CUG)** This corresponds to a group of users with limited possibilities of calling (only the people of the group and certain numbers).

■ **Operator-determined barring**   This provides restrictions of different services and call types by the operator.

GSM Architecture

A GSM network consists of several functional entities, whose functions and interfaces are defined. Figure 1-9 shows the layout of a generic GSM network. The GSM network can be divided into three broad parts. The subscriber carries the mobile station; the *Base Station Subsystem* (BSS) controls the radio link with the mobile station; and the Network Subsystem, the main part of which is the *Mobile services Switching Center* (MSC), performs the switching of calls between the mobile and other fixed or mobile network users, as well as the management of mobile services, such as authentication. The Operations and Maintenance Center, which oversees the proper operation and setup of the network, is not shown in the figure. The mobile station and the Base Station Subsystem communicate across the Um interface, also known as the air interface or radio link. The Base Station Subsystem communicates with the network service switching center across the A interface.

The added components of the GSM architecture (Figure 1-10) include the functions of the databases and messaging systems such a

■ *Home Location Register* (HLR)
■ *Visitor Location Register* (VLR)

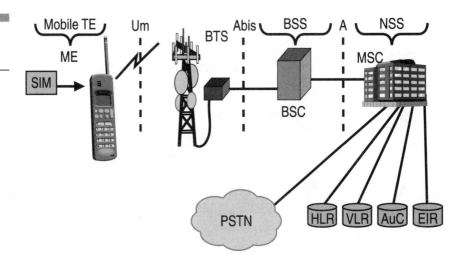

Figure 1-9
The GSM architecture.

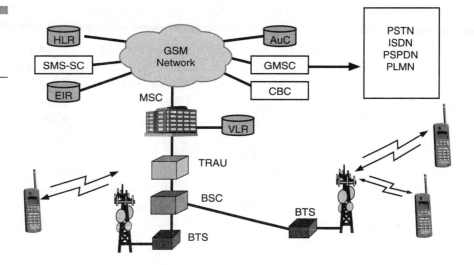

- *Equipment Identity Register* (EIR)
- *Authentication Center* (AuC)
- *SMS Serving Center* (SMS SC)
- *Gateway MSC* (GMSC)
- *Charge Back Center* (CBC)
- *Operations and Support Subsystem* (OSS)
- *Transcoder and Adaptation Unit* (TRAU)

Mobile Equipment or Mobile Station

The mobile station (MS) consists of the physical equipment, such as the radio transceiver, display and digital signal processors, and a smart card called the Subscriber Identity Module (SIM). It provides the air interface to the user in GSM networks. As such, other services are also provided, which include

- **Voice** Teleservices
- **Data** Bearer services
- **Features** Supplementary services

Subscriber Identity Module

The SIM provides personal mobility, so that the user can have access to all subscribed services irrespective of both the location of the terminal and the use of a specific terminal. By inserting the SIM card into another GSM cellular phone, as shown in Figure 1-11, the user is able to receive calls at that phone, make calls from that phone, or receive other subscribed services. The *International Mobile Equipment Identity* (IMEI) uniquely identifies the mobile equipment. The SIM card contains the *International Mobile Subscriber Identity* (IMSI), identifying the subscriber, a secret key for authentication, and other user information. The IMEI and the IMSI are independent, thereby providing personal mobility. A password or personal identity number may protect the SIM card against unauthorized use.

The Mobile Station Function

Different types of terminals are available that are distinguished principally by their power and application:

- The fixed terminals are terminals installed in cars.
- The GSM portable terminals can also be installed in vehicles.

Figure 1-11
The SIM.

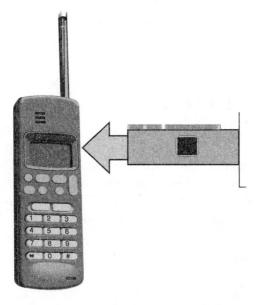

■ The hand-held terminals have experienced the biggest success thanks to their weight and volume, which are continuously decreasing.

The mobile station also provides the receptor for SMS messages, enabling the user to toggle between the voice and data use. Moreover, the mobile facilitates access to voice-messaging systems. The mobile station also provides access to the various data services available in a GSM network. These data services include

■ X.25 packet switching through a synchronous or asynchronous dial-up connection to the PAD at speeds typically at 9.6 Kbps

■ General Packet Radio Services using either an X.25- or IP-based data transfer method at speeds up to 115 Kbps

■ High-speed circuit-switched data at speeds up to 64 Kbps

The data speeds will vary by application and other conditions such as air interfaces across a hostile link.

The Base Transceiver Station (BTS)

The *Base Transceiver Station* (BTS) (Figure 1-12) houses the radio transceivers that define a cell and handles the radio link protocols with the mobile station. In a large urban area, a large number of BTSs may be deployed. The requirements for a BTS are

■ Ruggedness

■ Reliability

■ Portability

■ Minimum cost

The BTS corresponds to the transceivers and antennas used in each cell of the network. A BTS is usually placed in the center of a cell. Its transmitting power defines the size of a cell. Each BTS has between 1 and 16 transceivers depending on the density of users in the cell. Each BTS serves a single cell. It also includes the following functions:

■ Encoding, encrypting, multiplexing, modulating, and feeding the RF signals to the antenna

■ Transcoding and rate adaptation

■ Time and frequency synchronizing

Figure 1-12
The BTS.

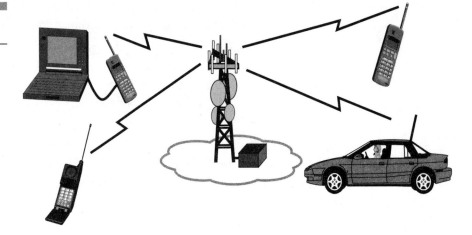

- Voice through full- or half-rate services
- Decoding, decrypting, and equalizing received signals
- Random access detection
- Timing advances
- Uplink channel measurements

The Base Station Controller (BSC)

The *Base Station Controller* (BSC) manages the radio resources for one or more BTSs. It handles radio channel setup, frequency hopping, and handovers. The BSC is the connection between the mobile and the Mobile service Switching Center (MSC). The BSC also translates the 13-Kbps voice channel used over the radio link to the standard 64-Kbps channel used by the *Public-Switched Telephone Network* (PSDN) or ISDN. The BSC is between the BTS and the MSC, and provides radio resource management for the cells under its control. It assigns and releases frequencies and time slots for the MS. The BSC also handles intercell handover. It controls the power transmission of the BSS and MS in its area. The function of the BSC is to allocate the necessary time slots between the BTS and the MSC. It is a switching device that handles the radio resources. Additional functions include

- Control of frequency hopping
- Performing traffic concentration to reduce the number of lines from the MSC
- Providing an interface to the Operations and Maintenance Center for the BSS
- Reallocation of frequencies among BTS
- Time and frequency synchronization
- Power management
- Time delay measurements of received signals from the mobile station

Base Station Subsystem

The Base Station Subsystem is composed of two parts: the Base Transceiver Station (BTS) and the Base Station Controller (BSC). These communicate across the specified Abis interface, enabling (as in the rest of the system) operation between components that are made by different suppliers. The radio components of a BSS may consist of four to seven or nine cells. A BSS may have one or more BS. The BSS uses the Abis interface between the BTS and the BSC. A separate high-speed line (T1 or E1) is then connected from the BSS to the Mobile Central Office, as shown in the architecture in Figure 1-13.

Figure 1-13
The Base Station
Subsystem.

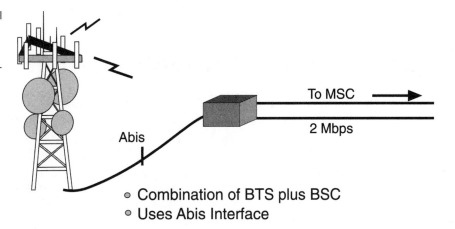

Abis

To MSC →

2 Mbps

- Combination of BTS plus BSC
- Uses Abis Interface

The Transcoder and Adaptation Unit (TRAU)

Depending on the costs of transmission facilities from a cellular operator, it may be cost efficient to have the transcoder either at the BTS, BSC, or MSC. If the transcoder is located at the MSC, it is functionally still a part of the BSS. This creates maximum flexibility of the overall network operation. The transcoder takes the 13-Kbps speech or data (at 300, 600, 1,200 bps) multiplexes four of them and places them on a standard 64-Kbps digital PCM channel. First, the 13-Kbps voice is brought up to a 16-Kbps level by inserting additional synchronizing data to make up the difference of the lower data rate. Then, four 16-Kbps channels are multiplexed onto a DS0 (64-Kbps) channel.

Locating the TRAU

If the transcoder/rate adapter is outside the BTS, the Abis interface can only operate at 16 Kbps within the BSS. The TRAU output data rate is 64-Kbps standard digital channel capacity. Next, 30 64-Kbps channels are multiplexed onto a 2.048-Mbps E1 service if the CEPT standards are used. The E1 can carry up to 120 traffic and control signals (16–120). The locations can be between the BTS and the BSC whereby a 16-Kbps subchannel is used between the BTS and the TRAU and 64-Kbps channels between the TRAU and the BSC. Alternatively, the TRAU can be located between the BSC and the MSC, as seen in Figure 1-14, using 16 Kbps at the BTS to BSC and 16 Kbps between the BSC and the TRAU.

Mobile Switching Center

The central component of the Network Subsystem is the Mobile services Switching Center (MSC), which is shown in Figure 1-15. It acts like a normal Class 5 *Central Office* (CO) in the PSTN or ISDN, and in addition provides all the functionality needed to handle a mobile subscriber, such as registration, authentication, location updating, handovers, and call routing to a roaming subscriber. The primary functions of the MSC include

- Paging
- Coordination of call setup for all MSs in its operating area

Figure 1-14
The TRAU.

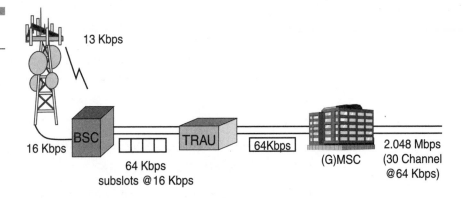

- In between BSC and MSC
- Converts GSM coding into PSTN data

13 Kbps ⟶ 64 Kbps   64 Kbps ⟶ 13 Kbps

Figure 1-15
The MSC.

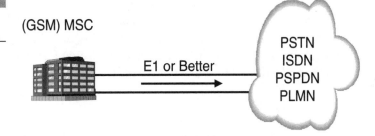

- Dynamic allocation of resources
- Location registration
- Interworking functions
- Handover management
- Billing
- Reallocation of frequencies to BTSs
- Encryption
- Echo cancellation
- Signaling exchange

- Synchronizing the BSS
- Gateway to SMS

As a CO function, it uses the digital trunks in the form of E1 (or larger) to the other network interfaces such as

- PSTN
- ISDN
- *Packet-Switched Public Data Network* (PSPDN)
- *Public Land Mobile Network* (PLMN)

These services are provided in conjunction with several functional entities, which together form the Network Subsystem. The MSC provides the connection to the public-fixed network (PSTN or ISDN), and signaling between functional entities uses *Signaling System Number 7* (SS7), used in ISDN and widely used in current public networks.

The *Gateway Mobile services Switching Center* (GMSC) is used in the PLMN. A gateway is a node interconnecting two networks. The GMSC is the interface between the mobile cellular network and the PSTN. It is in charge of routing calls from the fixed network towards a GSM user. The GMSC is often implemented in the same machines as the MSC. A PLMN may have many MSCs, but it has only one gateway access to the wireline network to accommodate the network operator. The gateway then is the high-speed trunking machine connected via E1 or *Synchronous Digital Hierarchy* (SDH) to the outside world.

The Registers Completing the NSS

The Home Location Register (HLR) and Visitor Location Register (VLR), together with the MSC, provide the call routing and roaming capabilities of GSM, called the *Network Switching Systems* (NSS). The HLR contains all the administrative information of each subscriber registered in the corresponding GSM network, along with the current location of the mobile. The current location of the mobile is in the form of a *Mobile Station Roaming Number* (MSRN), which is a regular ISDN number used to route a call to the MSC where the mobile is currently located. One HLR exists logically per GSM network, although it may be implemented as a distributed database. Figure 1-16 shows the HLR.

The VLR contains selected administrative information from the HLR, which is necessary for *call control* (CC) and provision of the subscribed

Figure 1-16
The HLR.

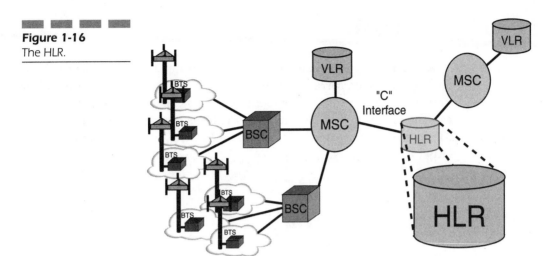

services, for each mobile currently located in the geographical area controlled by the VLR. Although each functional entity can be implemented as an independent unit, most manufacturers of switching equipment implement one VLR together with one MSC (Figure 1-17) so that the geographical area controlled by the MSC corresponds to that controlled by the VLR, simplifying the signaling required. Note that the MSC does not contain information about particular mobile stations—this information is stored in the location registers.

The other two registers are used for authentication and security purposes. The Equipment Identity Register (EIR) is a database that contains a list of all valid mobile equipment on the network, where its International Mobile Equipment Identity (IMEI) identifies each mobile station. An IMEI is marked as invalid if it has been reported stolen or is not type approved. The Authentication Center (AuC) is a protected database that stores a copy of the secret key stored in each subscriber's SIM card, which is used for authentication and ciphering of the radio channel.

Figure 1-17
The VLR.

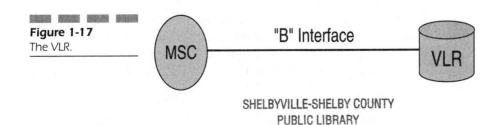

The Cell

As it has already been explained, a cell, identified by its *Cell Global Identity* (CGI) number, corresponds to the radio coverage of a base transceiver station. In a macrocell environment, the radius distance is between 3 to 35 km. The distances are calculated on the basis of a round-trip between the BTS and the mobile to provide sufficient *bit error rate* (BER) and power to satisfy quality speech.

Location Area

A *location area* (LA), identified by its *location area identity* (LAI) number, is a group of cells served by a single MSC/VLR. One MSC/VLR combination has several location areas. The LA is part of the MSC/VLR service area in which a mobile station may move freely without any updating of location messaging to the MSC/VLR controlling the location area.

MSC/VLR Service Area

A group of location areas under the control of the same MSC/VLR defines the MSC/VLR area. A single PLMN can have several MSC/VLR service areas. MSC/VLR is a sole controller of calls within its area of jurisdiction. To route a call to a mobile station, the path through the network links to the MSC in the MSC area where the subscriber is currently located. The mobile location can be uniquely identified because the MS is registered in a VLR, which is associated with an MSC.

Public Land Mobile Network (PLMN)

A Public Land Mobile Network (PLMN) is the area served by one network operator, as shown in Figure 1-18. One country can have several PLMNs, based on its size. The links between a GSM/PLMN network and other PSTN, ISDN, or PLMNs will be at the level of national or international

Figure 1-18
The PLMN.

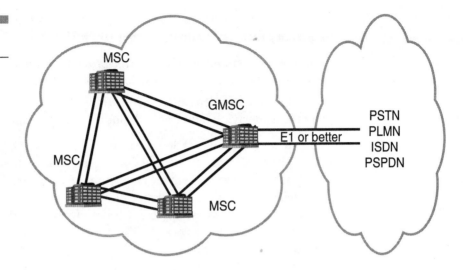

transit. All incoming calls for a GSM/PLMN will be routed to the Gateway MSC. A Gateway MSC works as an incoming transit exchange for the GSM/PLMN. All *mobile-terminated* (MT) calls will be routed to the Gateway MSC. Call connections between PLMNs or fixed networks must be routed through certain designated MSCs.

OSI Model—How GSM Signaling Functions in the OSI Model

The *Open Standards Interface* (OSI) is a guideline of how systems communicate transparently. SS7 is used for signaling between the outside world and the GSM architectures. Moreover, SS7 is used to communicate between the MSC and the HLR. To satisfy other functions in GSM architecture, the model is applied for other services from the mobile station outward. In reality, the model works at the bottom three layers of the OSI model for the bulk of the transmissions that take place in call setup and teardown, registration and authentication, and so on. Thus, Layers 3, 2, and 1 of the OSI model are most applicable.

OSI defines a communications subsystem consisting of functions that enable distributed application processes, resident on computers, to exchange information via an underlying data network. The communications subsystem can be divided into two sublayers:

■ An application-dependent sublayer providing functions that are application-dependent but network-independent

■ A network-dependent sublayer providing functions that are dependent on the underlying data network but are application-independent

Ensuring the transmission of voice or data of a given quality over the radio link is only part of the function of a cellular mobile network. A GSM mobile can seamlessly roam nationally and internationally, which requires that registration, authentication, call routing, and location-updating functions exist and are standardized in GSM networks. In addition, the fact that the geographical area covered by the network is divided into cells necessitates the implementation of a handover mechanism. The Network Subsystem performs these functions using the *Mobile Application Part* (MAP) built on top of the Signaling System Number 7 protocol, as shown in Figure 1-19.

Layer Functionality

In the GSM architecture, the layered model integrates and links the peer-to-peer communications between two different systems. If we look across the platform, the underlying layers satisfy the services of the upper-layer

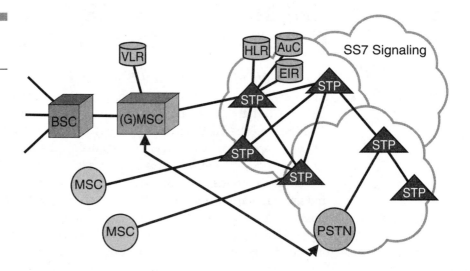

Figure 1-19
SS7 and GSM working together.

protocols. For example, at Layer 3, the *Service Access Point* (SAP) between Layer 3 and 2 addresses the services being served. *Service Access Point Identifiers* (SAPIs) describe the services that are provided by the various services from the upper and lower layers. Notifications are passed from layer to layer to ensure that the information has been properly formatted, transmitted, and received. These primitives make the process complete. Several discussions center on the chart of protocols, as shown in Figure 1-20. Refer to this chart for the next block of protocol stack discussions.

Mobile Station Protocols

The signaling protocol in GSM is structured into three general layers, depending on the interface. Layer 1 is the physical layer, which uses the channel structures discussed above over the air interface. Layer 2 is the data-link layer. Across the Um interface, the data-link layer is a modified version of the *Link Access Procedure for the D Channel* (LAPD) protocol used in ISDN, called *Link Access Protocol on Dm Channel* (LAPDm). Across the A interface, the *Message Transfer Part* (MTP) Layer 2 of Signaling System Number 7 is used. Layer 3 of the GSM signaling protocol is divided into three sublayers: *radio resources management* (RR), *mobility management* (MM), and *connection management* (CM).

Figure 1-20
The protocol stacks.

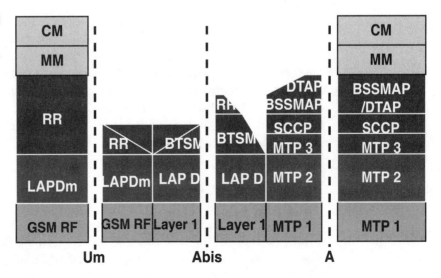

The Mobile Station to BTS Protocols

The radio resources (RR) management layer oversees the establishment of a link, both radio and fixed, between the mobile station and the MSC. The main functional components involved are the mobile station, the Base Station Subsystem, and the MSC. The RR layer is concerned with the management of an RR-session, which is the time that a mobile is in dedicated mode, as well as the configuration of radio channels including the allocation of dedicated channels.

The mobility management (MM) layer is built on top of the RR layer, and handles the functions that arise from the mobility of the subscriber, as well as the authentication and security aspects. Location management is concerned with the procedures that enable the system to know the current location of a powered-on mobile station so that incoming call routing can be completed.

Location updating a powered-on mobile is informed of an incoming call by a paging message sent over the PAGCH channel of a cell. One extreme would be to page every cell in the network for each call, which is obviously a waste of radio bandwidth. The other extreme would be for the mobile to notify the system, via location-updating messages, of its current location at the individual cell level. This would require paging messages to be sent to exactly one cell, but would be very wasteful due to the large number of location-updating messages. A compromised solution used in GSM is to group cells into location areas. Updating messages are required when moving between location areas, and mobile stations are paged in the cells of their current location area.

The connection management (CM) layer is responsible for call control (CC), supplementary service management, and short message service management. Each of these may be considered as a separate sublayer within the CM layer. Call control attempts to follow the ISDN procedures specified in Q.931, although routing to a roaming mobile subscriber is obviously unique to GSM. Other functions of the CC sublayer include call establishment, selection of the type of service (including alternating between services during a call), and call release.

BSC Protocols

After the information is passed from the BTS to the BSC, a different set of interfaces is used. The Abis interface is used between the BTS and BSC. At this level, the radio resources at the lower portion of Layer 3 are changed from the RR to the *Base Transceiver Station Management* (BTSM). The BTS management layer is a relay function at the BTS to the BSC. The RR protocols are responsible for the allocation and reallocation of traffic channels between the MS and the BTS. These services include controlling the initial access to the system, paging for mobile-terminated calls, handover of calls between cell sites, power control, and call termination. The RR protocols provide the procedures for the use, allocation, reallocation, and release of the GSM channels. The BSC still has some radio resource management in place for the frequency coordination, frequency allocation, and the management of the overall network layer for the Layer 2 interfaces.

From the BSC, the relay is using SS7 protocols so the MTP 1-3 is used as the underlying architecture and the BSS mobile application part or the Direct Application Part is used to communicate from the BSC to the MSC.

MSC Protocols

At the MSC, the information is mapped across the A interface to the MTP Layers 1 through 3 to the MSC from the BSC. Here the equivalent set of radio resources is called the BSS Mobile Application Part. The BSS *Mobile Application Part/Direct Termination Application Part* (MAP/DTAP) and the mobility management and connection management are at the upper layers of Layer 3 protocols. This completes the relay process. Through the control-signaling network, the MSCs interact to locate and connect to users throughout the network. Location registers are included in the MSC databases to assist in the role of determining how and whether connections are to be made to roaming users. Each user of a GSM MS is assigned a Home Location Register (HLR) that is used to contain the user's location and subscribed services. A separate register, the Visitor Location Register (VLR), is used to track the location of a user. As the users roam out of the area covered by the HLR, the MS notifies a new VLR of its whereabouts. The VLR in turn uses the control network (which happens to be based on SS7) to signal the HLR of the MS's new location. Through this information, mobile-terminated (MT) calls can be routed to the user by the location information contained in the user's HLR.

Defining the Channels

As we look at the radio operation, a channel can be defined in different ways. Often times we hear a channel defined in radio frequency. Other times we hear the physical channel being described thinking that is radio frequency. Alas, the different definitions get in the way. For the definitions used, channels are defined by looking at the matrix:

- Radio channel is defined by the frequency used.
- Physical channels are indicative of the time slot that they occupy.
- Logical channels are defined by the function that they provide or serve.

Frequencies Allocated

In reality, GSM systems can be implemented in any frequency band. However several bands exist where GSM terminals are available. Furthermore, GSM terminals may incorporate one or more of the GSM frequency bands listed in the following section to facilitate roaming on a global basis.

Two frequency bands, of 25 MHz in each one, have been allocated by ETSI for the GSM system:

- The band 890 to 915 MHz has been allocated for the uplink direction (transmitting from the mobile station to the base station).
- The band 935 to 960 MHz has been allocated for the downlink direction (transmitting from the base station to the mobile station).

However, not all countries can use all of the GSM frequency bands. This is due primarily to military reasons and to the existence of previous analog systems using part of the two 25-MHz frequency bands. Figure 1-21 shows the frequencies.

Primary GSM

When transmitting in a GSM network, the mobile station to the base station uses an uplink. The reverse channel direction is the downlink from the base station to the mobile station. GSM uses the circa 900-MHz band. The frequency band used is 890 to 915 MHz (mobile transmit) and 935 to 960 MHz (base transmit). The duplex channel enables the two-way communications in a GSM network. Because telephony was the primary service, a full-duplex channel is assigned with the two separate frequencies in a 45-MHz separation.

To give the maximum number of users access, each band is subdivided into 125 carrier frequencies spaced 200-kHz apart, using FDMA tech-

Figure 1-21
The uplink and downlink frequencies.

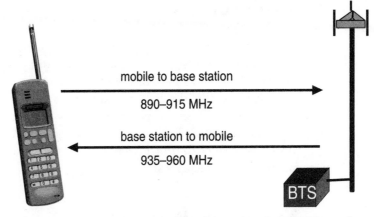

mobile to base station

890–915 MHz

base station to mobile

935–960 MHz

BTS

GSM frequencies initially set with 25 MHz (transmit and receive) spaced apart by 45 MHz.

Figure 1-22
Spectrum bands for primary GSM.

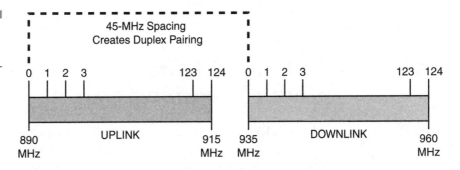

Channel 0 not used. Acts as guardband.

niques. The spectrum assignment is shown in Figure 1-22. Only 124 channels are used, where channel 0 is reserved and held as a guard band against interference from the lower channels. Each of these carrier frequencies is further subdivided into time slots using TDMA. The frequency bands are usually split between two or more providers who then build their networks. The channels are set at the 200 kHz each. The International Telecommunication Union (ITU), which manages the international allocation of radio spectrum (among other functions), allocated the bands for mobile networks in Europe.

Radio Assignment

Each BTS is assigned a group of channels with which to operate. Any frequency can be assigned to the BTS, as they are frequency agile. This enables the system to reallocate frequencies as needed to handle load balancing. Normally, the BTS can support upwards of 31 channels (frequencies); however, in actual operation, the operators usually assign from 1 to 16 channels per BTS. This is a business and practicality issue. The *Absolute Radio Frequency Channel Number* (ARFCN) is used in the channel assignment at each of the frequencies.

Frequency Pairing

The pairing is shown as the way of handling the 45-MHz separations. Remember that channel 0 was not used. It was reserved as a guard band from the lower frequencies to prevent interference.

Extended GSM Radio Frequencies

After the ETSI assigned the initial block of frequencies, a later innovation was to assign an additional block of 10 MHz on the bottom of the original block. The reasoning was that future demands would require this capacity. This meant that the frequencies from 880 to 890 MHz for the uplink and 915 to 925 MHz were added. This created an additional 50 carriers. The carriers were numbered 974 to 1,023 so that the channel assignments would not be confused with the initial GSM standard. Once the added channels were implemented, the additional channels were still paired at 45-MHz separation.

- Channel 974 was not used; it became the guard band for the lower frequencies below 880 MHz and 925 MHz.
- The initial channel 0 in the primary GSM band is now used because of this shift, as shown in Figure 1-23.

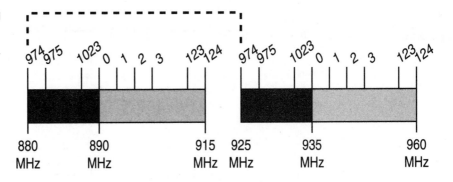

Figure 1-23
Extended GSM.

—Channel 974 not used. Acts as guardband.

—Channel 0 is now used.

Modulation

In order to convey the speech on the radio frequency, either in analog or digital form, the transmitted information must be propagated on the radio link. It must be placed on the carrier. A carrier in this respect is a single radio frequency. The process of combining the audio and the radio signals is known as modulation. The resultant waveform is known as a modulated waveform. Modulation is a form of change process where we change the input information into a suitable format for the transmission medium. We also unchanged the information by demodulating the signal.

Three normal forms of modulation are used:

- Amplitude
- Frequency
- Phase

Amplitude Shift Keying (ASK)

In *amplitude shift keying* (ASK) (Figure 1-24), the radio wave is modulated by shifting on the amplitude. The frequency is left constant, but the amplitude is shifted high if the data is a 1 and low if the data is a 0. Normally, we see two amplitude shifts represent a single bit.

Figure 1-24
ASK.

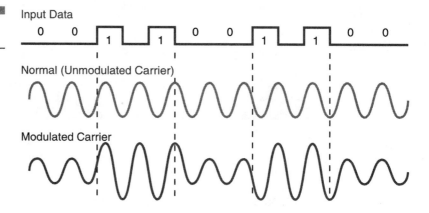

Figure 1-24
ASK.

Frequency Shift Keying (FSK)

The alternative to ASK is *frequency shift keying* (FSK). In the case of FSK (Figure 1-25), applying the data onto the radio wave modulates the carrier by changing the frequency. The amplitude is kept constant and the frequency is changed. Normally, a single frequency shift represents a bit of data.

Phase Shift Keying (PSK)

In *phase shift keying* (PSK), both the amplitude and the frequency are kept constant, so the changes are represented by a shift in the phase, as shown

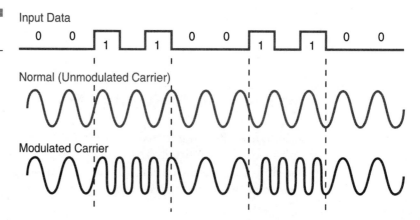

Figure 1-25
FSK.

Figure 1-26
PSK.

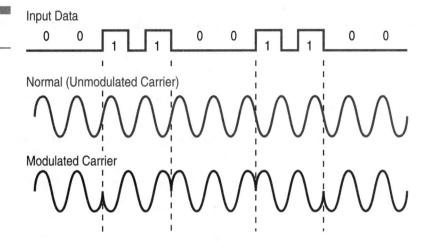

in Figure 1-26. The benefit of phase shifts is that multiple phases can be used to represent more than one modulated bit. Under normal PSK, a shift in the phase represents a single bit; however, multiphase modulation enables multiple bits to be represented. The quadrature phase (QPSK) shifts will allow up to 2 bits per shift, whereas a quadrature and amplitude shift will allow 4 bits per phase shift.

Gaussian Minimum Shift Keying (GMSK)

GSM modulation works differently, as seen in Figure 1-27. Using *Gaussian minimum shift keying* (GMSK), the nature of the data moved from the mobile station is digital. For a digital transmission in GSM, the chosen modulation scheme needs to have good error performance in light of the noise and interference in a mobile network environment. GMSK is a complex scheme based largely on mathematical functions. The basis of this scheme is an *offset quadrature phase shift keying* (OQPSK), which offers the advantage of a fairly narrow spectral output. This is combined with a minimum technique that controls the rate of change of the phase of the carrier and the radiated spectrum will be even lower. This also requires very careful planning at the sites to prevent interference and produces only 1 bit per symbol. The combined functions of the baseband filter, the OQPSK and GMSK modulation work to produce a compact transmission spectrum. This is important if adequate adjacent channel interference figures are to be met. The total symbol rate for GSM at 1 bit per symbol in GMSK produces

Figure 1-27
GMSK results.

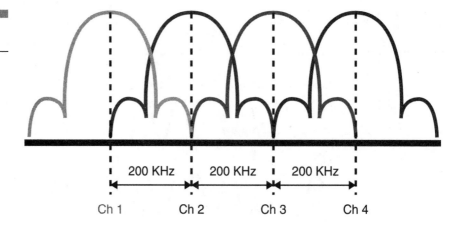

270.833 ksymbols/second. The gross transmission rate of the time slot is 22.8 Kbps.

Access Methods

Because radio spectrum is a limited resource shared by all users, a method must be devised to divide up the bandwidth among as many users as possible. The choices are

- Frequency Division Multiple Access (FDMA)
- Time Division Multiple Access (TDMA)
- *Code Division Multiple Access* (CDMA)

GSM chose a combination of Time and Frequency Division Multiple Access (TDMA/FDMA) as its method. The FDMA part involves the division by frequency of the total 25-MHz bandwidth into 124 carrier frequencies of 200-kHz bandwidth. One or more carrier frequencies are then assigned to each base station. Each of these carrier frequencies is then divided in time, using a TDMA scheme, into eight time slots. One time slot is used for transmission by the mobile and one for reception. They are separated in time so that the mobile unit does not receive and transmit at the same time, a fact that simplifies the electronics.

FDMA

The FDMA part involves the division by frequency of the total 25-MHz bandwidth into 124 carrier frequencies of 200-kHz bandwidth. One or more carrier frequencies are then assigned to each base station. Using FDMA, a frequency is assigned to a user, as seen in Figure 1-28. Therefore, the larger the number of users in an FDMA system, the larger the number of available frequencies must be. The limited available radio spectrum and the fact that a user will not free its assigned frequency until he or she does not need it anymore explain why the number of users in an FDMA system can be quickly limited.

TDMA

Time Division Multiple Access (TDMA) is digital transmission technology that enables a number of users to access a single radio frequency (RF) channel without interference by allocating unique time slots to each user within each channel. The TDMA digital transmission scheme multiplexes three signals over a single channel. Each of the carrier frequencies is divided in time, using a TDMA scheme, into eight time slots, as shown in Figure 1-29. One time slot is used for transmission by the mobile and one for reception. They are separated in time so that the mobile unit does not receive and transmit at the same time, a fact that simplifies the electronics.

TDMA enables several users to share the same channel. Each of the users, sharing the common channel, is assigned his or her own burst

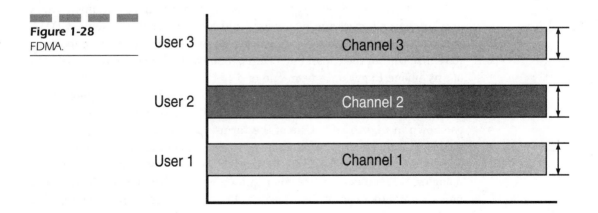

Figure 1-28
FDMA.

Figure 1-29
TDMA.

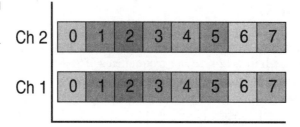

within a group of bursts called a frame. Usually, TDMA is used with an FDMA structure. In addition to increasing the efficiency of transmission, TDMA offers a number of other advantages over standard cellular technologies. First and foremost, it can be easily adapted to the transmission of data as well as voice communication. TDMA offers the capability to carry data rates of 64 Kbps to 120 Mbps (expandable in multiples of 64 Kbps). This enables operators to offer personal communication-like services including fax, voice band data, and Short Message Services (SMSs) as well as bandwidth-intensive applications such as multimedia and videoconferencing. Unlike spread-spectrum techniques that can suffer from interference among the users all of whom are on the same frequency band and transmitting at the same time, TDMA's technology, which separates users in time, ensures that they will not experience interference from other simultaneous transmissions.

CDMA

CDMA is characterized by high capacity and small cell radius, which employs spread-spectrum technology and a special coding scheme. CDMA is the dynamic allocation of bandwidth. To understand this, it's important to realize that in the context of CDMA, "bandwidth" refers to the capability of any phone to get data from one end to the other. It doesn't refer to the amount of spectrum used by the phone, because in CDMA every phone uses the entire spectrum of its carrier whenever it is transmitting or receiving, as shown in Figure 1-30. One of the terms you'll hear in conjunction with CDMA is "soft handoff." A handoff occurs in *any* cellular system when your call switches from one cell site to another as you travel. In all other technologies, this handoff occurs when the network informs your phone of the new channel to which it must switch. The phone then stops receiving and

Figure 1-30
CDMA.

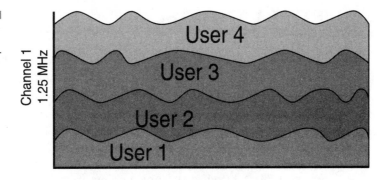

transmitting on the old channel, and commences transmitting and receiving on the new channel. It goes without saying that this is known as a hard handoff.

In CDMA, however, every phone and every site are on the same frequency. In order to begin listening to a new site, the phone only needs to change the pseudo-random sequence it uses to decode the desired data from the jumble of bits sent for everyone else. While a call is in progress, the network chooses two or more alternate sites that it feels are handoff candidates. It simultaneously broadcasts a copy of your call on each of these sites. Your phone can then pick and choose between the different sources for your call, and move between them whenever it feels like it. It can even combine the data received from two different sites to ease the transition from one to the other. CDMA is more efficient about that kind of thing. In both TDMA and CDMA, the outgoing voice traffic is digitized and compressed. However, the CDMA codec can realize when the particular packet is noticeably simpler (for example, silence or a sustained tone with little change in modulation) and will compress the packet far more. Thus, the packet may involve fewer bits, and the phone will take less time to transmit it. That's where this odd idea of what bandwidth means in CDMA comes in. For in a real sense, bandwidth in CDMA equates to receive power at the cell. CDMA systems constantly adjust power to make sure as little is used as necessary, and compensate for this by using coding gain through the use of forward error correction and other approaches that are much too complicated to go into. The chip rate is constant, and if more actual data is carried by the constant chip rate, then less coding gain will occur. Therefore, it's necessary to use more power instead.

TDMA Frames

In GSM, a 25-MHz frequency band is divided, using an FDMA scheme, into 124 carrier frequencies spaced one from each other by a 200-kHz frequency band. Normally, a 25-MHz frequency band can provide 125 carrier frequencies, but the first carrier frequency is used as a guard band between GSM and other services working on lower frequencies. Each carrier frequency is then divided in time using a TDMA scheme. This scheme splits the radio channel, with a width of 200 kHz, into eight bursts, as shown in Figure 1-31. A burst is the unit of time in a TDMA system, and it lasts approximately 0.577 ms. A TDMA frame is formed with eight bursts and lasts, consequently, 4.615 ms. Each of the eight bursts that form a TDMA frame are then assigned to a single user.

Time Slot Use

One time slot is used for transmission by the mobile and one for reception. They are separated in time so that the mobile unit does not receive and transmit at the same time, a fact that simplifies the electronics. A separation is used with a three-time slot offset so that the mobile will not have to send and receive at the same time.

GSM FDMA/TDMA Combination

To enable multiple access, GSM utilizes a blending of FDMA and TDMA. This combination is used to overcome the problems introduced in each indi-

Figure 1-31
TDMA framing and
time slots.

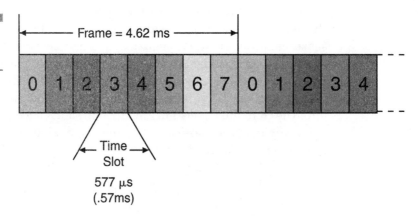

vidual scheme. In the case of FDMA, frequencies are divided up into smaller ranges of frequency slots and each of these slots is assigned to a user during a call. Although this method will result in an increase of the number of users, it is not efficient in the case of high user demand. On the other hand, TDMA assigns a time slot for each user for utilizing the entire frequency. Similarly, this will become easily overloaded when encountering high user demand. Hence, GSM uses a two-dimensional access scheme. GSM uses the combined FDMA and TDMA architecture to provide the most efficient operation within the scope of price and reasonable data. The physical channels are TDMA time slots and the radio channels are frequencies. This scheme divides the entire frequency bandwidth into several smaller pieces as in FDMA and each of these frequency slots is to be divided into eight time slots in a full-rate configuration. Similarly, 16 time slots will be in a half-rate configuration.

Logical Channels

GSM distinguishes between physical channels (the time slot) and logical channels (the information carried by the physical channels). Several recurring time slots on a carrier constitute a physical channel, which are used by different logical channels to transfer information—both user data and signaling. A channel corresponds to the recurrence of one burst every frame. It is defined by its frequency and the position of its corresponding burst within a TDMA frame. GSM has two types of channels:

- The traffic channels used to transport speech and data information
- The control channels used for network management messages and some channel maintenance tasks

The Physical Layer

Each physical channel supports a number of logical channels used for user traffic and signaling. The physical layer (or Layer 1) supports the functions required for the transmission of bit streams on the air interface. Layer 1 also provides access capabilities to upper layers. The physical layer is described in the GSM Recommendation 05 series (part of the ETSI documentation for GSM). At the physical level, most signaling messages carried on the radio path are in 23-octet blocks. The data-link layer functions are multiplexing, error detection and correction, flow control, and segmentation to enable long messages on the upper layers.

The radio interface uses the Link Access Protocol on Dm channel (LAPDm). This protocol is based on the principles of the ISDN Link Access Protocol on the D channel (LAPD) protocol. Layer 2 is described in GSM Recommendations 04.05 and 04.06. The following logical channel types are supported:

- *Speech traffic channels* (TCHs)
  - *Full-rate TCH* (TCH/F)
  - *Half-rate TCH* (TCH/H)
- *Broadcast channels* (BCHs)
  - *Frequency correction channel* (FCCH)
  - *Synchronization channel* (SCH)
  - *Broadcast control channel* (BCCH)
- *Common control channels* (CCCHs)
  - *Paging channel* (PCH)
  - *Random access channel* (RACH)
  - *Access grant channel* (AGCH)
- *Cell broadcast channel* (CBCH)
  - *Cell broadcast channel* (CBCH) (the CBCH uses the same physical channel as the DCCH)
- *Dedicated control channels* (DCCHs)
  - *Slow associated control channel* (SACCH)
  - *Stand-alone dedicated control channel* (SDCCH)
  - *Fast associated control channel* (FACCH)

Speech Coding on the Radio Link

The transmission of speech is, at the moment, the most important service of a mobile cellular system. The GSM speech codec (coder and decoder), which will transform the analog signal (voice) into a digital representation, has to meet the following criteria:

- It must have good speech quality, at least as good as the quality obtained with previous cellular systems.

- Reduce the redundancy in the sounds of the voice. This reduction is essential due to the limited capacity of transmission of a radio channel.

- The speech codec must not be very complex because complexity is equivalent to high costs.

The final choice for the GSM speech codec is a codec named *Regular Pulse Excitation Long-Term Prediction* (RPE-LTP). This codec uses the information from previous samples (this information does not change very quickly) in order to predict the current sample. The speech signal is divided into blocks of 20 ms; these blocks are then passed to the speech codec, which has a rate of 13 Kbps, in order to obtain blocks of 260 bits.

Channel Coding

Channel coding adds redundancy bits to the original information in order to detect and correct, if possible, errors that occurred during the transmission. The channel coding is performed using two codes: a block code and a convolutional code.

- The block code corresponds to the block code defined in the GSM Recommendations 05.03. The block code receives an input block of 240 bits and adds four zero tail bits at the end of the input block. The output of the block code is consequently a block of 244 bits.

- A convolutional code adds redundancy bits in order to protect the information. A convolutional encoder contains memory. This property differentiates a convolutional code from a block code. A convolutional code can be defined by three variables: n, k, and K. The value n corresponds to the number of bits at the output of the encoder, k to the number of bits at the input of the block, and K to the memory of the encoder.

Convolutional Coding

Before applying the channel coding, the 260 bits of a GSM speech frame are divided in three different classes according to their function and importance. The most important class is the class Ia containing 50 bits. The class Ib is next in importance, which contains 132 bits. The class II is the least important, which contains the remaining 78 bits. The different classes are coded differently. First of all, the class Ia bits are block coded. Three parity bits, used for error detection, are added to the 50 class Ia bits. The resultant

53 bits are added to the class Ib bits. Four zero bits are added to this block of 185 bits (50 + 3 + 132). A convolutional code, with $r = \frac{1}{2}$ and $K = 5$, is then applied, obtaining an output block of 378 bits. The class II bits are added, without any protection, to the output block of the convolutional coder. An output block of 456 bits is finally obtained.

This description is meant to set the stage for understanding the underlying network that supports the GPRS systems. Much more detail and descriptive materials can be found in other publications, so that the reader can gain a better description if needed. In the next chapter, the focus will be on the motivators for the operators to move from a GSM-only network to one that overlays GPRS on the GSM network. In each of the following chapters, we will look at the interfaces between each of the components in a GSM/GPRS network.

GPRS
Introduction

Objectives

When you complete the reading in this chapter, you will be able to

- Describe the main objectives for moving to GPRS.
- Understand when the services will be available for commercial rollout.
- Discuss the choices of terminals available.
- Understand applications for use with GPRS.
- Describe the radio interfaces for data applications.
- Discuss the speeds we can expect to achieve.

Introduction to the Internet and Wireless Wave

With the simultaneous gate-opening effects of technological innovation and industry deregulation, the demand for communications and available solutions is exploding. This demand is being fueled by the needs of people and businesses. The most visible evidence of the boom is within Internet traffic and e-commerce or m-commerce. However, it is less appreciated that an unprecedented demand exists from worldwide telephone subscribers. It took a century to get 700 million phone lines installed. Another 700 million will be deployed in the next 15 to 20 years—and that could prove to be a conservative estimate.

Although the majority of the new deployments will be wireless phones—700 million of them over the next 10 years—demand for wireline communications is also exploding, driven in part by the need to access the Internet. This explosion in demand is reflective of the dependence that people have on rapid, reliable communications to keep up with the fast pace of business.

The success of *Global Systems for Mobile* (GSM), the ubiquitous presence it has garnered, the emerging Internet, and the overall growth of data traffic in general all point to a significant business opportunity for GSM operators. The number of subscribers to the Internet worldwide is growing exponentially, as seen in Figure 2-1, and the growth has been dramatic. The following statistics from the middle of 2001 add some credibility to the overall concept of a data-centric community that is also mobile.

- Number of Internet users—400 million
- Number of wireless users—700 million

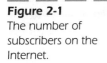

Figure 2-1
The number of
subscribers on the
Internet.

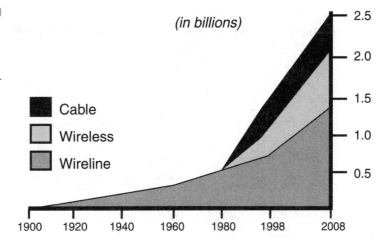

By the year 2005, we can expect that more than 1 billion people will be mobile wireless users on a worldwide basis and by 2006, more than 1 billion people will be Internet users. We are adding over 100 million new Internet users per year. This growth is unheralded in the past century of the telecommunications industry. The phenomenal growth in wireline Internet subscribers points to the possibility of wireless operators capturing some of this market. This assumes that they can offer comparable price-performance capabilities. The factors that we can consider in the process are

■ Wireless users are ideal target subscribers for Internet providers.

■ Internet users are ideal target subscribers for GSM operators.

Several movements in the Internet community will also force changes. The demographics of the user will change dramatically as the world expands its wireless and Internet presence. The average user will be looking for developments in the new Internet that will provide a broader band communications speed, the capability of using a network-enabled appliance, and a full network capable of sourcing all the needs and applications through high-end portals that can offer the goods and services needed. Instant information for a mobile workforce is paramount.

In addition to customer demand, this revolution is greatly accelerated by technology disruptions. Three technologies are at the fundamental science level. One is summarized in the well-known Moore's Law, which states that the capacity of an individual chip will double once every 18 to 24 months. Therefore, silicon is covering the globe.

That phenomenon has been going on for three decades now and will be adding as much capacity in the next two years as has been created in the history of the semiconductor business. Nevertheless, the capacity is finally just getting to the point where it's interesting.

Two other technologies, although less well known, are changing just as rapidly, if not more so:

- In optical, in the core of the network, *Dense Wave Division Multiplexing* (DWDM), using multiple colors of light to send multiple data streams down the same optical fiber, is disrupting the rapid growth of *Wave Division Multiplexing* (WDM), further pushing the envelope.

- Wireless capacity is also exploding, enabling higher bandwidth for voice/data without fiber (45 Mbps—up to 2.5 Gbps with certain wireless tools).

Combined, these advances are making converged networks possible and inevitable, as well as important to plan for in business. Companies that understand and take advantage of this convergence will have a strategic advantage.

The New Wave of Internet User

During the next few years, the third-generation Internet will drive even further innovation and performance. Figure 2-2 provides a summary of the steps.

It began with the first-generation Internet, which was PC-driven. During this period, standards were established and narrowband services were offered. This led to business model experimentation, new companies, and new brands. In both wireline and wireless communications, users were satisfied with the PC-centric services because the networks did not offer anything else. This led to a somewhat frustrated PC and Internet user on wireline networks, but an even greater level of frustration was evident in the wireless arena.

Today, the second-generation Internet is upon us. Trends underlying today's Internet include substantial personalization and the emergence of the business-to-business market. Regardless of the downturn in 2001, the networks will reemerge.

Trends underlying the third-generation Internet will drive the growth of the new economy in upcoming years.

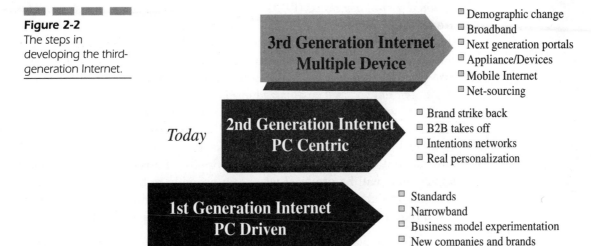

Figure 2-2
The steps in developing the third-generation Internet.

3rd Generation Internet
Multiple Device

- Demographic change
- Broadband
- Next generation portals
- Appliance/Devices
- Mobile Internet
- Net-sourcing

Today

2nd Generation Internet
PC Centric

- Brand strike back
- B2B takes off
- Intentions networks
- Real personalization

1st Generation Internet
PC Driven

- Standards
- Narrowband
- Business model experimentation
- New companies and brands

- First, as more people go online, in fact almost doubling, the online population will begin to normalize, or resemble the overall U.S. population. The average age of the online user will go up and his or her income will fall. This means that strong brand recognition increases in importance, greater service levels to support less savvy online users will be required, and convenience and ease of use will become vital. The good news is that these changes will drive a greater comfort level with online shopping for the average online user, more than doubling total spending online.

- Second, broadband will create a personalized, interactive experience. Think of the "Web on steroids." Today's interactive experience will be radically enhanced.

  - Instead of instant text messaging, we will have easy access to instant audio and video messaging.

  - Instead of today's chat rooms and discussion lists, we will have far more sophisticated real-time collaboration tools.

  - Today's grainy-streamed audio and video will have broadcast quality tomorrow. The two-dimensional will be three-dimensional.

Although the growth of PCs has slowed to roughly 3 percent per year, new information appliances and communication devices are fast becoming the new power brokers with double-digit growth. Wireless telephones, *personal digital assistants* (PDAs), Blackberry devices, and GPRS terminals are the new devices to behold. The power of the Web will be accessed through mobile phones, PDAs, cable set-top boxes, and even game controllers. Every office and household device in the future will be Internet Protocol (IP) addressable, enabling the user and supplier to better service the individualized and customized needs.

This new economic model (although it appears to have gradually slowed) requires that we address the converged world, borrowing from each perspective. Both the world of voice networks and the world of data networks have advantages. Both have a unique profile of strengths. Convergence applications will be practical when you are able to take the best of both worlds and deliver real-world business value. In a nutshell, a converged world requires that our networks and access methods of the future be

- Highly reliable
- Broadband-serviced
- Scaleable
- Multiservice-oriented
- Flexible and open
- Exceptionally easy to use

These thrusts are driving the data communications market into an explosive situation. The average growth of our voice networks is 4 percent; however, in the data communications arena, the growth is still approximating 30 percent growth per year. This unparalleled growth consists of both goods and services to meet the demands of customers, internal users, and the industry in general.

General Packet Radio Service (GPRS)

General Packet Radio Service (GPRS) is a key milestone for GSM data. It offers end users new data services and enables operators to offer radically new pricing options. Using the existing GSM radio infrastructure, up-front investments for operators are relatively low. GPRS solutions began appearing initially in 1999 through 2000 using the infrastructures that are already in place. Pricing for use of the voice side of the network has become commoditized, whereas pricing models for the new data access will set a new revolution. One such threshold looks at an all-you-can-eat model whereby users of wireless phones add a data subscription at $29.95 per month for unlimited use. Another such model is the one used in Japan by DoCoMo by charging a rate of the U.S. equivalent of $.0025 per packet. Others will emerge that will shake the industry mode and create new dynamics in the use of data anywhere.

GPRS services were targeted at the business user. However, the services will soon be available networkwide, targeting both the business and the residential consumer. The widespread adoption and acceptance of GPRS will create a critical mass of users, driving down costs while offering better services. These components will form the basis of a healthy mobile data market with growth figures comparable with GSM voice-only services today.

Research by Infonetics indicates that the movement of the user community will also be to a more mobile community. In fact, the study indicates that by 2005, more wireless devices will be used for the Internet than PCs on the Net, as shown in Figure 2-3. This form of growth is again a driver that will force the rapid deployment by carriers and manufacturers alike.

The GPRS Story

The GPRS is a new service that provides actual packet radio access for GSM and *Time Division Multiple Access* (TDMA) users alike. The main benefits of GPRS are that it reserves radio resources only when data is

Figure 2-3
Within five years, more wireless devices will be used than PCs on the Internet.

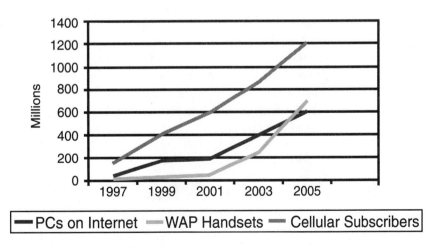

Figure 2-4
The GPRS story.

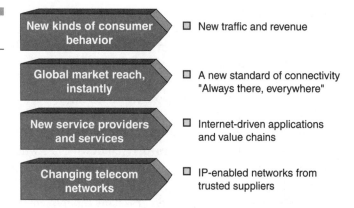

available to send, and it reduces the reliance on traditional circuit-switched networks. Figure 2-4 is a basic stepping-stone description of the GPRS story. The increased functionality expected from GPRS will decrease the incremental costs to provide the data services, which will increase the penetration of the data services, among the consumers and business users alike.

Additionally, GPRS will improve the quality of data services measured in terms of reliability, response time, and features available. Unique applications will be developed in the future that will attract the broad base of mobile users and enable the individual providers to offer differentiated services. One way that GPRS improves upon the capacity capabilities of the network suppliers is to share the same radio resources among all mobile stations in a cell, thus providing effective use of scarce resources. New core network elements will continue to emerge that will expand services, features, and operations for our bursty data applications.

GPRS also provides an added step toward *third-generation* (3G) networks. GPRS will enable the network operators to implement IP-based core architecture for data applications. This will continue to proliferate new services and mark the steps to 3G services for integrated voice and data applications.

What Is GPRS?

As stated previously, GPRS stands for General (or generic) Packet Radio Service. GPRS extends the packet data capabilities of the GSM networks from *Packet Data on Signaling-channel Service* (PDSS) to higher data rates and longer messages. For now, the use of GPRS shall be in the context of GPRS-GSM to distinguish it from the GPRS-136, the North American adoption of GPRS by the IS-136-based systems. GPRS is designed to coexist with the existing GSM *Public Land Mobile* Network (PLMN). It may be deployed as an overlay onto the existing GSM radio network. GPRS may also be implemented incrementally in specific geographic areas. An example of this GPRS radio access may be deployed in some cells of a GSM network, but perhaps not all. As the demand grows, coverage can be expanded. A network view of GPRS is shown generically in Figure 2-5.

The GPRS network fits in with the existing GSM PLMN as well as the existing packet data networks. GPRS PLMN provides the wireless access to the wired packet data networks. GPRS shares resources between packet data services and other services on the GSM PLMN. GPRS PLMN also interworks with the *Short Message Service* (SMS) components to provide SMS over GPRS. The intent is to provide a seamless network infrastructure for operations and maintenance of the network.

GPRS is a packet-based data bearer service for GSM and TDMA (IS-136) networks, which provides both standards with a way to handle higher-data speeds and the transition to 3G. It will make mobile data faster, cheaper, and user-friendlier than ever before. By introducing packet switching and *Internet Protocol* (IP) to mobile networks, GPRS gives mobile users faster data speeds, and particularly suits bursty Internet and intranet traffic. For the subscriber, GPRS enables voice and data calls to be handled simultaneously. Connection setup is almost instantaneous, and users can have

Figure 2-5
A GPRS network view.

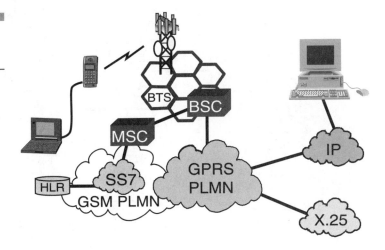

always-on connectivity to the mobile Internet, enjoying high-speed delivery of e-mails with large file attachments, Web surfing, and access to corporate LANs.

GPRS was defined by the *European Telecommunications Standards Institute* (ETSI) as a means of providing a true packet radio service on GSM networks. GSM equipment vendors are actively developing systems that adhere to the GPRS specifications. At the same time, carriers whose networks are based on *North American TDMA* (NA-TDMA) (IS-136) have decided to deploy GPRS technologies in their networks. Internetworking and interoperability specifications have been developed between ANSI/IS-136 and GSM; therefore, this is a logical extension of the overall scheme. Figure 2-6 is an example of the internetworking arrangements that are planned for use within GPRS.

This creates a coup for the ETSI, because up to now, IS-136 networks have been completely based on *Telecommunications Industry Association* (TIA) standards and specifications. Today, GPRS is seen as one of the preliminary steps down a path that will someday lead to the convergence of GSM and IS-136 networks.

Figure 2-6
Internetworking
strategies in GPRS.

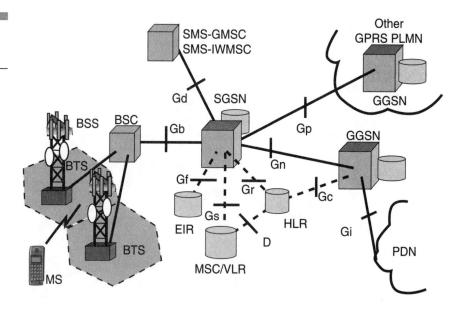

Market Timeline for GPRS

The deployment timeline for GPRS is dependent on several factors, including the infrastructure availability and terminal availability. An overall deployment timeline shows the starting point in the first half of 2000 for the GPRS infrastructure. Initial availability of infrastructure and terminal devices includes use for trials and limited-scale deployment (controlled roll-out). General availability refers to availability on a widespread commercial deployment to the masses.

In addition to the GPRS timeline, it will be necessary to investigate the 3G deployments because they are closely related. Because the GPRS operators are planning to deploy 3G (or at least are looking at it), GPRS is the migration step toward 3G. Several proof-of-concept-type trials have been underway for some time. These trials have led to more technical and application-oriented trials in the latter part of 2001 and into 2002. As with GPRS, terminal and infrastructure availability are the driving factors. Moreover, completion of licensing processes is necessary for commercial deployment.

Motivation for GPRS

GPRS was developed to enable GSM operators to meet the growing demands for wireless packet data service that is a result of the explosive growth of the Internet and corporate intranets. Applications using these networks require relatively high throughput and are characterized by bursty traffic patterns and asymmetrical throughput needs. Applications such as Web browsing typically result in bursts of network traffic while information is being transmitted or received, followed by long idle periods while the data is being viewed. In addition, much more information is usually flowing to the client device than is being sent from the client device to the server. GPRS systems are better suited to meet the demand of this bursty data need than the traditional circuit-switched wireless data systems. GPRS allocates the bandwidth independently in the uplink and downlink.

Another goal for GPRS is to enable GSM operators to enter the wireless packet data market in a cost-efficient manner. First, they must be able to provide data services without changing their entire infrastructure. The initial GPRS standards make use of standard GSM radio systems. This also includes GSM standard modulation schemes and TDMA framing structures. By doing this, the cost implications are minimized in the cell equipment. Second, GSM operators must have flexibility to deploy GPRS without having to commit their entire network to it. GPRS provides the dynamic allocation and assignment of radio channels to packet services according to the demand.

Evolution of Wireless Data

Data support over *first-generation* (1G) wireless networks started with *Advanced Mobile Phone System* (AMPS), circuit-switched data communications, as shown in the graph in Figure 2-7. This worked by attaching a cellular modem (a standard modem that supports the AMPS wireless interface) with a laptop computer. This began the evolution to the first wireless packet data networks—*Cellular Digital Packet Data* (CDPD), with data rates up to 19.2 Kbps, as shown in the graph in Figure 2-8. CDPD works with AMPS networks and was initially designed for short intermittent transactions, such as credit card verification, e-mail, and fleet dispatch services. According to the Wireless Data Forum, CDPD covered 55 percent of the U.S. population as early as the *third quarter of 1998* (3Q98). It has since grown to cover nearly 87 percent based on the proliferation of more

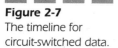

Figure 2-7
The timeline for circuit-switched data.

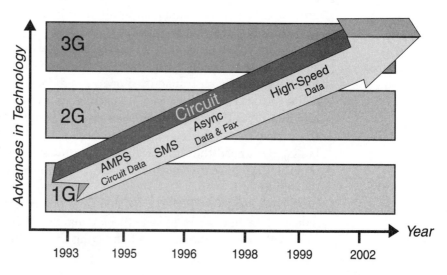

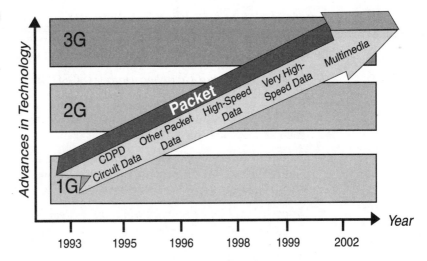

Figure 2-8
The timeline for
packet-switched data.

wireless users. In addition, limited support for SMS was introduced to offer
paging-like and text-messaging services.

In *second-generation* (2G) wireless networks, SMS services became the
deployed architecture of choice. It uses the existing infrastructure of 2G
wireless networks, with the addition of the Message Center component. 2G
also introduced asynchronous data and facsimile services over the air interfaces, with initial data rates of up to 14.4 Kbps. This enables users to fax
and have dial-up access to an ISP account, corporate account, and the like.
Packet data technology gained momentum in 2G and then on to 2.5G networks. This includes GPRS and packet data support in *Code Division Multiple Access* (CDMA). Data rates for the packet switching currently range
from 9.6 to 19.2 Kbps. In the future of 2G, we can expect to see data rates
at up to 115 Kbps.

3G, when it happens, will support data rates of 384 Kbps to 2 Mbps. Multimedia and high-speed Internet access will be the expected normalized
data access applications.

Wireless Data Technology Options

Today, GSM has the capability to handle messages via the SMS and
14.4-Kbps circuit-switched data services for data and fax calls. The maximum speed of 14.4 Kbps is relatively slow compared to the wireline modem

speeds of 33.6 and 56 Kbps. To enhance the current data capabilities of GSM, operators and infrastructure providers have specified new extensions to GSM Phase II, as shown in Figure 2-9, to provide

- *High-Speed Circuit-Switched Data* (HSCSD) by using several circuit channels

- GPRS to provide packet radio access to external packet data networks (such as X.25 or Internet)

- *Enhanced Data rate for GSM Evolution* (EDGE) using a new modulation scheme to provide up to three times higher throughput (for HSCSD and GPRS)

- *Universal Mobile Telecommunication System* (UMTS), a new wireless technology using new infrastructure deployment

These extensions enable

- Higher data throughput

- Better spectral efficiency

- Lower call setup times

The way to implement GPRS is to add new packet data nodes in GSM/TDMA networks and upgrade existing nodes to provide a routing

Figure 2-9
Steps of
implementation.

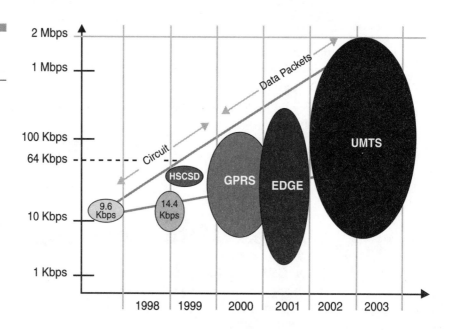

path for packet data between the mobile terminal and a gateway node. The gateway node will provide interworking with external packet data networks for access to the Internet and intranets, for example. Few or no hardware upgrades are needed in existing GSM/TDMA nodes, and the same transmission links will be used between Base Transceiver Stations and Base Station Controllers for both GSM/TDMA and GPRS.

GPRS Roaming

At the end of June 2001, 551 million GSM customers were on record and 447 operators in 170 countries have now adopted GSM. The expectation is that GSM growth will continue and it will have 700 million GSM customers by June 2002. As the growth continues and more people disconnect from their wired phones, GSM will have 800 million customers by the end of 2003 and 1 billion customers by 2005. Obviously, this places a lot of growth on the providers' networks and puts more demand on the ability to use the service wherever and whenever we want. GSM Voice Roaming was a 15 billion Euro business in 1999, thus indicating that the masses are using their phones today in a roaming manner. Some statistics about the wireless roaming environment in Europe are as follows:

- 540 million roaming calls were made in February 2000.
- 750 million calls were predicted in June, July, and August 2000.
- Data will account for between 20 and 50 percent of all global wireless traffic by 2004.
- 8 billion short messages were sent in May 2000.
- 10 billion SMS messages were sent in December 2000.
- 1 billion SMS messages were sent per month in Europe alone in 2001.
- GSM grew at 80 percent in 1999; PCs grew at 22 percent.
- All terminals will be Internet-enabled by 2002 to 2003.
- More GSM terminals will be connected to the Internet than PCs by 2005.
- Wireless devices will access 30 percent of all Internet traffic by 2005.

The GSM Phase II Overlay Network

The typical GPRS PLMN enables a mobile user to roam within a geographic coverage area and receive continuous wireless packet data services. The user may move while actively sending and receiving data or may move during periods of inactivity. Either way, the network tracks the location of the mobile station so incoming packets can be routed to the mobile station when they arrive. The GPRS PLMN interfaces with the mobile stations via the air interface. GPRS services will initially be provided using an enhanced version of the standard GSM interface. The operators will evolve their networks to incorporate more advanced radio interfaces in the future so that they can deliver higher data rates to the end user.

The GPRS PLMN interfaces as an overlay to traditional public packet data networks using standard *Packet Data Protocols* (PDPs), as shown in Figure 2-10. The network layer protocols supported for interfacing with packet data networks include X.25 and the IP. Through these networks, the

Figure 2-10
The GPRS overlay on GSM.

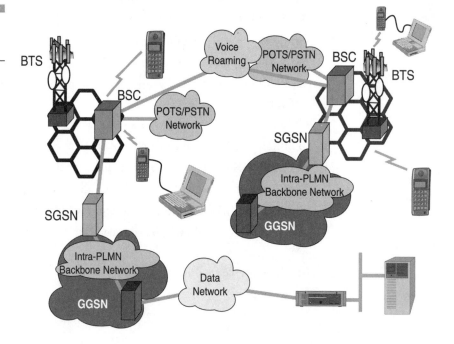

end user is able to access public servers such as the Web sites and private corporate intranet servers. GPRS can also receive voice services via the GSM PLMN. Voice services and GPRS services may be accessed alternately or simultaneously depending on the mobile station's capabilities. Several classes of mobile stations are possible, which vary in degree of complexity and capability. The actual end-user data terminal used can be a smart phone, a dedicated wireless data terminal, or a standard data terminal connected to a GPRS-capable phone.

Circuit-Switched or Packet-Switched Traffic

An On/Off model characterizes the typical Internet data. The user spends a certain amount of time downloading Web pages in quick succession, followed an indeterminate long time of inactivity during which he or she may be reading the information, thinking, or maybe even have left the work space. In fact, the traffic is quite bursty (sporadic) and can be characterized as data packets averaging about 16 Kbps in size with average inter-arrival times of about seven seconds. If a circuit-switched connection is used to access the Internet, then the bandwidth that is dedicated for the entire duration of the session is underutilized. This inefficient use of the circuit-switched example shown in Figure 2-11 creates an undesirable scenario for the network operators. Instead, they would like to fill the channels (circuits) to the highest reasonable level and carry as much billable traffic as possible.

GPRS involves overlaying a packet-based air interface on the existing circuit-switched GSM network shown in Figure 2-12. This gives the user an option to use a packet-based data service. To supplement a circuit-switched network architecture with packet switching is quite a major upgrade.

However, the GPRS standard is delivered in a very elegant manner—with network operators needing only to add a couple of new infrastructure nodes and make a software upgrade to some of the existing network elements. With GPRS, the information is split into separate, but related packets before being transmitted and reassembled at the receiving end. Packet switching is similar to a jigsaw puzzle—the image that the puzzle represents is divided into pieces at the factory where it is made and then the pieces are placed into a plastic bag. During the transport of the new-boxed

Figure 2-11
The circuit-switched
traffic example.

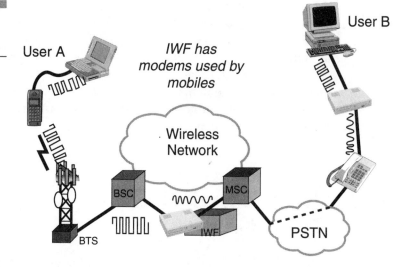

Figure 2-11
The circuit-switched
traffic example.

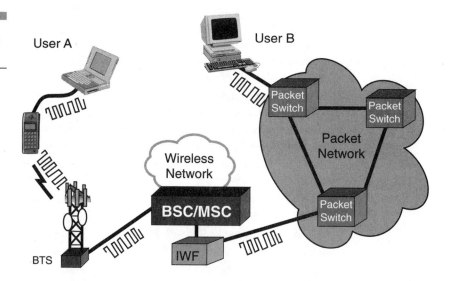

Figure 2-12
The packet-switched
example.

puzzle from the factory to the end user, the pieces get all mixed up. When the final recipient receives the bag, all the pieces are reassembled into the original image. All the pieces are related and fit together, but the way they are transported and reassembled varies by system, as seen in Figure 2-13. The Internet is another example of this type of a packet data network.

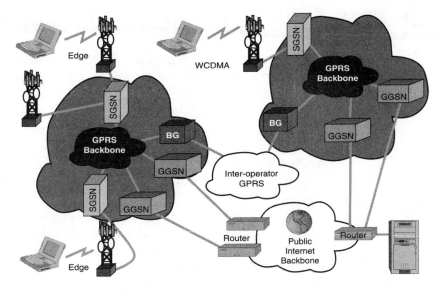

Figure 2-13
The pieces of GPRS
and GSM fit together.

GPRS Radio Technologies

Packet switching means that the GPRS radio resources are used only when users are actually sending or receiving data. Rather than dedicating a radio channel to one mobile user for a fixed period of time, the available radio resources can be concurrently shared by several users. This efficient use of the scarce radio resources means that a larger number of GPRS users can share the same bandwidth and be served from a single cell. The actual number of users supported depends on the application being used and how much data each user has to send or receive. Because the spectrum efficiency is improved in GPRS, it is not as necessary to build idle capacity that is only used during peak transmit hours. GPRS therefore lets the operator maximize system usage and efficiency in a dynamic and flexible way.

In fact, all eight time slots of a TDMA frame can be made available to each user. However, as the number of simultaneous users increases, collisions will occur between the randomly arriving data packets. This will cause queuing delays on the downlink. Therefore, the effective throughput perceived by each user decreases, but more gracefully. The idea of concatenation or aggregation of the time slots to be available to one user makes this far more palatable for the end user to understand how he or she can bundle services together and run the data faster.

Cells and Routing Areas

The geographic coverage area of a GPRS network is divided into smaller areas known as cells and routing areas, as shown in Figure 2-14.

A cell is the area that is served by a set of radio base stations. When a GPRS mobile station wants to send data or prepare to receive data, it searches for the strongest radio signal that it can find. Once the mobile scans for the strongest signal and locates the strongest base station, it then notifies the network of the cell it is receiving the strongest and selects it. At this point, the mobile listens to the base station for news of incoming data packets.

Periodically, the mobile station uses its idle time to listen to transmissions from neighboring base stations and evaluates the signal quality of their transmissions. If the mobile determines that a different base station signal is received stronger (better) than the current base, then the mobile may begin to listen to the new base station instead. This means that the mobile will listen to a different signal. The process of moving from one base station to another is called cell reselection. In some cases, the mobile station informs the network that it has changed cells by performing a location update procedure.

When data arrives for an idle mobile station, the network broadcasts a notice that it wants to establish communications with that mobile. This is called paging and is very similar to the paging process in wireless voice networks.

Figure 2-14
Cells and routing areas.

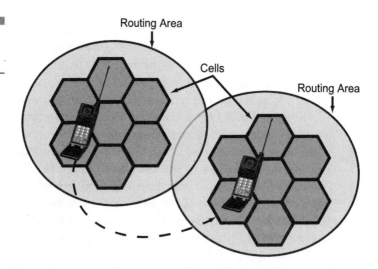

A group of neighboring cells can be grouped together to form a routing area. Network engineers use routing areas to strike a balance between location-updating traffic and paging traffic. Mobile stations that have been actively sending or receiving data are tracked at the cell level. (The network keeps track of the cell that they are currently using.) Mobile stations that have been inactive (idle) are tracked at the routing area level (the network keeps track of the routing area).

Attaching to the Serving GPRS Support Node

When a GPRS mobile station wants to use the wireless packet data network services, it must first attach to a *Serving GPRS Support Node* (SGSN), as shown in Figure 2-15. When the SGSN receives a request from a mobile station, it makes sure that it wants to honor the request for service. Several factors must first be considered:

■ Is the mobile a subscriber to GPRS services? The act of verifying the mobile station's subscription information is called *authorization*.

■ Is the mobile who it says it is? The process of verifying the identity of the mobile station is called *authentication*.

Figure 2-15
Attaching to the SGSN.

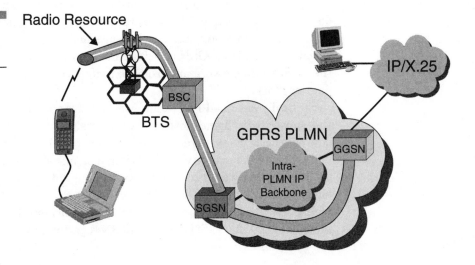

■ What level of *quality of service* (QoS) is the station requesting? Has the mobile subscribed to the level of QoS being requested (is the owner willing to pay for it?) and can the network provide this level of service while still providing the levels of service already promised to the other attached users?

Once the SGSN decides to accept an attachment, it keeps track of the mobile station as the mobile moves around in the coverage area. The SGSN needs to know where the mobile is in case data packets arrive and need to be routed to the mobile. Attaching to the SGSN is similar to creating a logical connection (or pipe) between the mobile and the SGSN. The logical connection is maintained as the mobile moves within the coverage area controlled by the SGSN.

The attachment to an SGSN is not sufficient to enable the mobile to begin transferring packet data. To do that, the mobile needs to activate (and possibly acquire) a PDP address (such as an IP address).

Packet Data Protocol (PDP) Contexts

The PDP addresses are network layer addresses (*Open Standards Interconnect* [OSI] model Layer 3). GPRS systems support both X.25 and IP network layer protocols. Therefore, PDP addresses can be X.25, IP, or both. Each PDP address is anchored at a *Gateway GPRS Support Node* (GGSN), as shown in Figure 2-16. All packet data traffic sent from the public packet data network for the PDP address goes through the gateway (GGSN). The public packet data network is only concerned that the address belongs to a specific GGSN. The GGSN hides the mobility of the station from the rest of the packet data network and from computers connected to the public packet data network.

Statically assigned PDP addresses are usually anchored at a GGSN in the subscriber's home network. Conversely, dynamically assigned PDP addresses can be anchored either in the subscriber's home network or the network that the user is visiting. When a mobile station is already attached to a SGSN and wants to begin transferring data, it must activate a PDP address. Activating a PDP address establishes an association between the mobile's current SGSN and the GGSN that anchors the PDP address. The record kept by the SGSN and the GGSN regarding this association is called the PDP context.

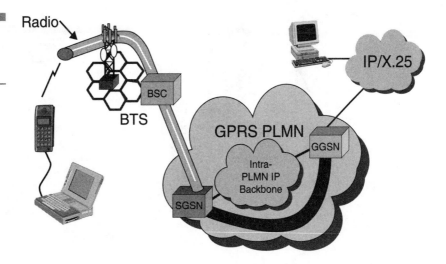

Figure 2-16
Obtaining a PDP
context from the
GGSN.

It is important to understand the difference between a mobile station attaching to a SGSN and a mobile station activating a PDP address. A single mobile station attaches to only one SGSN; however, it may have multiple PDP addresses that are all active at the same time. Each of the addresses may be anchored to a different GGSN.

If packets arrive from the public packet data network at a GGSN for a specific PDP address and the GGSN does not have an active PDP context corresponding to that address, it may simply discard the packets. Conversely, the GGSN may attempt to activate a PDP context with a mobile station if the address is statically assigned to a particular mobile.

Data Transfer

Once the mobile station has attached to a SGSN and activated a PDP address, it is now ready to begin communicating with other devices. For example, a GPRS mobile is communicating with a computer system connected to an X.25 or IP network. The other computer may be unaware that the mobile station is, in fact, mobile. It may only know the mobile station's PDP address. The packets, as shown in Figure 2-17, need to be routed as follows:

Assume that the mobile station has attached to an SGSN and activated its PDP address. Packets sent from the other computer to the mobile station

Figure 2-17
The data transfer.

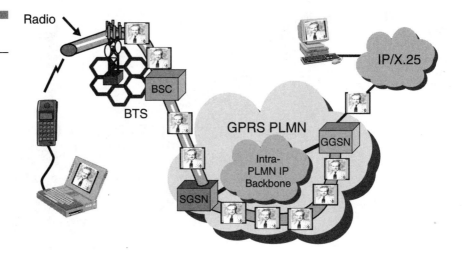

first travel across the public packet data network to reach the GGSN that anchors the PDP address. From here, the GGSN must forward the packets to the SGSN to which the mobile station is currently attached. Obviously, packets flowing in the reverse direction must be first routed through the SGSN and GGSN before being passed to the public packet data network.

Communication between the *GPRS Serving Nodes* (GSNs) makes use of a technique known as tunneling. Tunneling is the process of wrapping the network layer packets into another header so that they can be routed through the GPRS PLMN IP backbone network. Inside the network, packets are routed based on the new header alone and the original packet is carried as the payload. Once they reach the far side of the GPRS network, they are unwrapped and continue along their way through the external network. From this point onward, they are routed based on their original (internal) header. Using tunneling within GPRS solves the mobility problem for the packet networks and helps to eliminate the complex task of protocol interworking.

Mobile IP also makes use of tunneling to route packets to mobile nodes. In mobile IP, packets are only tunneled from the fixed network to the mobile station. Packets flowing from the mobile to fixed nodes use normal routing. GPRS, by contrast, uses tunneling in both directions.

GSM and NA-TDMA Evolution

Both GSM and NA-TDMA are evolving from 2G to 3G and GPRS plays an important role in this evolution. Some of the key points to note are as follows:

- As evident, attempts are made to seek synergy between the two TDMA base systems now (as opposed to what happened 10 years ago).

- GSM-GPRS standards and concepts are being adopted in North American TDMA as GPRS-136. The radio interface is being adapted to 30-kHz channels and an IS-136 DCCH channel structure. In fact, many of the North American carriers (AT&T Wireless and Voice Stream, among others) are planning to offer GPRS on GSM as an evolution from the NA-TDMA architecture).

- EDGE has adopted the eight-PSK-modulation scheme that is used for 136+.

- 136HS and EDGE are being developed with synergy in mind. UWC-136 has embraced the GPRS/EDGE architecture for 200-kHz-wide 136HS Outdoors.

- The North American and European proposals differ for the 2.0-Mbps systems. UWC-136 continues to use a purely TDMA scheme, whereas CDMA-based UTRA is the *Radio Transmission Technology* (RTT) of choice for ETSI.

GPRS Terminals

GPRS terminals can be grouped in three GPRS mobile station classes, each having different capabilities to fulfill market needs:

- **Class A**   A mobile station that can make/receive calls on both GSM and GPRS simultaneously.

- **Class B**   The mobile station can make and/or receive calls on either GSM or GPRS but not simultaneously.

- **Class C**   The mobile station can be either in GPRS or in GSM mode (manually selected).

GPRS will drive the convergence of mobile computing and wireless. In addition, it is expected that apart from the traditional wireless terminal vendors such as Motorola, Nokia, Ericsson, Panasonic, and Mitsubishi, mobile computing device vendors like Sony and PDA vendors like 3COM and Handspring will enter the GPRS terminal market. Class A mobile devices are the most complex and lag behind the other classes.

Mobile Station Classes for Multislot Capabilities

GPRS multislot class refers to the different capabilities to transmit and receive on different combinations of multiple time slots. Twenty-nine different classes are available:

Class 1 = One receive and one transmit slot

Class 29 = Eight transmit and eight receive slots

The class defines the number of time slots allowed for the uplink and the number of time slots for the downlink. Both the downlink and the uplink can be different due to the nature of nonsymmetrical traffic.

- Type 1 mobile stations are not required to transmit and receive at the same time.
- Type 2 mobile stations are required to transmit and receive at the same time.

R_x describes the maximum number of receive time slots that the mobile station can use per TDMA frame. The mobile must be able to support all integer values of receive time slots from zero to R_x (depending on the services supported by the mobile station). The receive time slot need not be contiguous. For Type 1 mobile stations, the receive time slots shall be allocated within the window of size R_x, and no transmit time slots shall occur between receive time slots within a TDMA frame.

T_x describes the maximum number of transmit time slots that the mobile station can use per TDMA frame. The mobile station must be able to support all integer values of transmit time slots from zero to T_x (depending on the services supported by the mobile station). The transmit time slots need not be contiguous. For Type 1 mobile stations, the transmit time slots shall be allocated within the window of size T_x, and no receive time slots shall occur between transmit time slots within a TDMA frame.

Table 2-1 shows the types of multislot terminals and the way they have been categorized. The 29 different classes and options are shown in this table.

Table 2-1

Twenty-nine
Classes of Mobile
Terminals

Multislot Class	Maximum Number of Slots			Minimum Number of Slots				Type
	R_x	T_x	Sum	T_{ta}	T_{tb}	T_{ra}	T_{rb}	
1	1	1	2	3	2	4	2	1
2	2	1	3	3	2	3	1	1
3	2	2	3	3	2	3	1	1
4	3	1	4	3	1	3	1	1
5	2	2	4	3	1	3	1	1
6	3	2	4	3	1	3	1	1
7	3	3	4	3	1	3	1	1
8	4	1	5	3	1	2	1	1
9	3	2	5	3	1	2	1	1
10	4	2	5	3	1	2	1	1
11	4	3	5	3	1	2	1	1
12	4	4	5	2	1	2	1	1
13	3	3	NA	NA	a	3	a	2
14	4	4	NA	NA	a	3	a	2
15	5	5	NA	NA	a	3	a	2
16	6	6	NA	NA	a	2	a	2
17	7	7	NA	NA	a	1	a	2
18	8	8	NA	NA	0	0	0	2
19	6	2	NA	3	b	2	c	1
20	6	3	NA	3	b	2	c	1
21	6	4	NA	3	b	2	c	1
22	6	4	NA	2	b	2	c	1
23	6	6	NA	2	b	2	c	1
24	8	2	NA	3	b	2	c	1
25	8	3	NA	3	b	2	c	1
26	8	4	NA	3	b	2	c	1
27	8	4	NA	2	b	2	c	1
28	8	6	NA	2	b	2	c	1
29	8	8	NA	2	b	2	c	1

a = 1 with frequency hopping
  0 without frequency hopping

b = 1 with frequency hopping or changing R_x to T_x
  0 without frequency hopping and no changing R_x to T_x

c = 1 with frequency hopping or changing from T_x to R_x
  0 without frequency hopping and no changing from T_x to R_x

Applications for GPRS

Many applications fit into the mode of GPRS and IPs. These applications are merely a means to an end. In other scenarios, the features and applications can be met with other technologies. The issue at hand is that the use of GPRS facilitates these applications and drives the acceptance ratio. It is easy to say that we can do anything with GPRS, but it is more practical to say at a minimum that we can do the following.

Chat

Chat can be distinguished from general information services because the source of the information is a person with the chat protocol, whereas it tends to be from an Internet site for information services. The information intensity, the amount of information transferred per message, tends to be lower with chat, where people are more likely to state opinions than factual data. In the same way as Internet chat groups have proven to be a very popular application of the Internet, groups of like-minded people, so-called communities of interest, have begun to use nonvoice mobile services as a means to chat and discuss.

Because of its synergy with the Internet, GPRS would enable mobile users to participate fully in existing Internet chat groups rather than needing to set up their own groups that are dedicated to mobile users. Because the number of participants is an important factor determining the value of participation in the news group, the use of GPRS here would be advantageous. GPRS will not, however, support point-to-multipoint services in its first phase, hindering the distribution of a single message to a group of people. As such, given the installed base of SMS-capable devices, we would expect SMS to remain the primary bearer for chat applications in the foreseeable future, although experimentation with using GPRS is likely to commence sooner rather than later.

Textual and Visual Information

A wide range of content can be delivered to mobile phone users ranging from share prices, sports scores, weather, flight information, news headlines, prayer reminders, lottery results, jokes, horoscopes, traffic, location-sensitive services, and so on. This information does not necessarily need to be textual—it may be maps or graphs or other types of visual information.

The length of a short message of 160 characters suffices for delivering information when it is quantitative, such as a share price or a sports score or temperature. When the information is of a qualitative nature, however, such as a horoscope or news story, 160 characters is too short other than to tantalize or annoy the information recipient because they receive the headline or forecast but little else of substance. As such, GPRS will likely be used for qualitative information services when end users have GPRS-capable devices, but SMS will continue to be used for delivering most quantitative information services. Interestingly, chat applications are a form of qualitative information that may remain delivered using SMS, in order to limit people to brevity and reduce the incidence of spurious and irrelevant posts to the mailing list that are a common occurrence on Internet chat groups.

Still Images

Still images such as photographs, pictures, postcards, greeting cards, presentations, and static Web pages can be sent and received over the mobile network as they are across fixed telephone networks. It will be possible with GPRS to post images from a digital camera connected to a GPRS radio device directly to an Internet site, enabling near real-time desktop publishing.

Moving Images

Over time, the nature and form of mobile communication is getting less textual and more visual. The wireless industry is moving from text messages to icons, picture messages to photographs, blueprints to video messages, movie previews being downloaded, and on to full-blown movie watching via data streaming on a mobile device.

Sending moving images in a mobile environment has several vertical market applications including monitoring parking lots or building sites for intruders or thieves, and sending images of patients from an ambulance to a hospital. Videoconferencing applications, in which teams of distributed salespeople can have a regular sales meeting without having to go to a particular physical location, are another application for moving images.

Web Browsing

Using circuit-switched data for Web browsing has never been an enduring application for mobile users. Because of the slow speed of circuit-switched data, it takes a long time for data to arrive from the Internet server to the browser. Alternatively, users switch off the images, just access the text on the Web, and end up with text layouts on screens that are difficult to read. As such, mobile Internet browsing is better suited to GPRS.

Document Sharing/Collaborative Working

Mobile data facilitates document sharing and remote collaborative working. This lets different people in different places work on the same document at the same time. Multimedia applications combining voice, text, pictures, and images can even be envisaged. These kinds of applications could be useful in any problem-solving exercise such as fire fighting, combat (to plan the route of attack), medical treatment, advertising copy setting, architecture, journalism, and so on. This collaborative working environment can be useful anytime a user can benefit from having the ability to comment on a visual depiction of a situation or matter. By providing sufficient bandwidth, GPRS facilitates multimedia applications such as document sharing.

Audio

Despite many improvements in the quality of voice calls on mobile networks such as *Enhanced Full Rate* (EFR), they are still not broadcast quality. In some scenarios, journalists or undercover police officers with portable professional broadcast-quality microphones and amplifiers capture interviews with people or radio reports that they have dictated and need to send this information back to their radio or police station. Leaving a mobile phone on, or dictating to a mobile phone, would not give sufficient voice quality to enable that transmission to be broadcast or analyzed for the purposes of background noise analysis or voice printing, where the speech autograph is taken and matched against those in police storage. Because even short voice clips occupy large file sizes, GPRS or other high-speed mobile data services are needed.

Job Dispatch

Nonvoice mobile services can be used to assign and communicate new jobs from office-based staff to mobile field staff. Customers typically telephone a call center whose staff takes the call and categorizes it. Those calls requiring a visit by a field sales or service representative can then be escalated to those mobile workers. Job dispatch applications can optionally be combined with vehicle-positioning applications, so that the nearest available suitable personnel can be deployed to serve a customer. GSM nonvoice services can be used not only to send the job out, but also as a means for the service engineer or salesperson to keep the office informed of progress towards meeting the customer's requirement. The remote worker can send in a status message such as "Job 1234 complete, on my way to 1235."

The 160 characters of a short message are sufficient for communicating most delivery addresses such as those needed for a sale, service, or some other job dispatch application such as mobile pizza delivery and courier package delivery. However, the 160 characters require manipulation of the customer data such as the use of abbreviations such as "St" instead of "Street." The 160 characters do not leave much space for giving the field representative any information about the problem that has been reported or the customer profile. The field representative is able to arrive at the customer premises but is not very well briefed beyond that. This is where GPRS will be beneficial to enable more information to be sent and received more easily. With GPRS, a photograph of the customer and his or her premises could, for example, be sent to the field representative to assist in finding and identifying the customer. As such, we expect job dispatch applications will be an early adopter of GPRS-based communications.

Corporate E-mail

With up to half of employees typically away from their desks at any one time, it is important for them to keep in touch with the office by extending the use of corporate e-mail systems beyond an employee's office PC. Corporate e-mail systems run on *Local Area Networks* (LAN) and include Microsoft Mail, Outlook, Outlook Express, Microsoft Exchange, Lotus Notes, and Lotus cc:Mail.

Because GPRS-capable devices will be more widespread in corporations than among the general mobile phone user community, more corporate e-mail applications are likely to use GPRS than Internet e-mail applications whose target market is more general.

Internet E-mail

Internet e-mail services come in the form of a gateway service where the messages are not stored, or mailbox services in which messages are stored. In the case of gateway services, the wireless e-mail platform translates the message from SMTP, the Internet e-mail protocol, into SMS and sends it to the SMS Center. In the case of mailbox e-mail services, the e-mails are actually stored and the user receives a notification on his or her mobile phone and can then retrieve the full e-mail by dialing in to collect it, forward it, and so on.

Upon receiving a new e-mail, most Internet e-mail users are not currently notified of this fact on their mobile phone. When they are out of the office, they have to dial in speculatively and periodically to check their mailbox contents. However, by linking Internet e-mail with an alert mechanism such as SMS or GPRS, users can be notified when a new e-mail is received.

Vehicle Positioning

This application integrates satellite-positioning systems that tell people where they are with nonvoice mobile services that enable people to tell others where they are. The *Global Positioning System* (GPS) is a free-to-use global network of 24 satellites run by the U.S. Department of Defense. Anyone with a GPS receiver can receive his or her satellite position and thereby find out where he or she is. Vehicle-positioning applications can be used to deliver several services including remote vehicle diagnostics, ad hoc stolen vehicle tracking, and new rental car fleet tariffs.

The SMS is ideal for sending GPS position information such as longitude, latitude, bearing, and altitude. GPS coordinates are typically about 60 characters in length. GPRS could alternatively be used.

Remote LAN Access

When mobile workers are away from their desks, they clearly need to connect to the LAN in their office. Remote LAN applications encompass access to any applications that an employee would use when sitting at his or her desk, such as access to the intranet, his or her corporate e-mail services such as Microsoft Exchange or Lotus Notes, and to database applications running on Oracle or Sybase. The mobile terminal such as a hand-held or laptop computer has the same software programs as the desktop on it, or cut-down client versions of the applications accessible through the corporate LAN. This application area is therefore likely to be a conglomeration of remote access to several different information types—e-mail, intranet, databases. This information may all be accessible through Web browsing tools, or require proprietary software applications on the mobile device. The ideal bearer for remote LAN access depends on the amount of data being transmitted, but the speed and latency of GPRS make it ideal.

File Transfer

As this generic term suggests, file transfer applications encompass any form of downloading sizeable data across the mobile network. This data could be a presentation document for a traveling salesperson, an appliance manual for a service engineer, or a software application such as Adobe Acrobat Reader to read documents. The source of this information could be one of the Internet communication methods such as *File Transfer Protocol* (FTP), telnet, http, or Java, or from a proprietary database or legacy platform. Irrespective of the source and type of file being transferred, this kind of application tends to be bandwidth-intensive. Therefore, it requires a high-speed mobile data service such as GPRS, EDGE, or UMTS to run satisfactorily across a mobile network.

Home Automation

Home automation applications combine remote security with remote control. Basically, you can monitor your home from anywhere—on the road, on vacation, or at the office. If your burglar alarm goes off, not only are you alerted, but also you can go live and see live footage of the perpetrators. You can program your video or switch on your oven so that the preheating is complete by the time you arrive home (traffic jams permitting). Your GPRS-capable mobile phone really becomes the remote control device for our television, video, and stereo. Because the IP will soon be everywhere, these devices can be addressed and fed instructions. A key enabler for home automation applications will be Bluetooth, which enables disparate devices to interwork.

These features and the driving motivators will propel the operators into the implementation of GPRS. Moreover, the applications will offer many new opportunities to users that were heretofore unavailable. It is no wonder that the hype of GPRS is strong now. The next approach we will look to will be the architecture of the GPRS infrastructure. This will help the reader to understand the overall architectural model used for GPRS.

System Architecture

Objectives

When you complete the reading in this chapter, you will be able to

■ Describe the main architecture of a GPRS network.

■ Describe the new components needed to operate a GPRS overlay.

■ Understand how the mobility management works.

■ Discuss the role of the gateways (signaling, charging, and IP routers).

■ Understand the different logical packet channels.

■ Describe all the different interfaces in GPRS.

Network Architecture

Support of *General Packet Radio Service* (GPRS) does not represent a major upgrade to the existing *Global Systems for Mobile* (GSM) infrastructure. The greatest impact is the addition of two new network elements, which are shown in Figure 3-1.

■ The *Serving GPRS Support Node* (SGSN)

■ The *Gateway GPRS Support Node* (GGSN)

Figure 3-1
The SGSN and GGSN
additions.

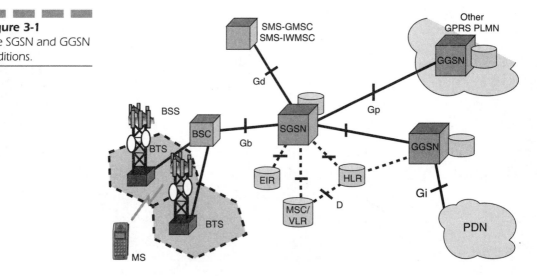

Functionally, no hardware impact occurs to the *Base Transceiver Systems* (BTS). Overall, GPRS represents a software upgrade to the *Base Station System* (BSS), with the exception of the introduction of *Packet Control Unit Support Nodes* (PCUSN) to support the packet orientation of the G_b interface logically between the *Base Station Controller* (BSC) and the SGSN.

The architecture of GPRS is designed so that signaling and high-level data protocols are system-independent. Only the low-level protocols in the radio interface must be changed to operate with the same services.

The SGSN can be viewed as a packet-switched *Mobile Switching Center* (MSC); it delivers packets to mobile stations within its service area. SGSNs send queries to *Home Location Registers* (HLRs) to obtain profile data of GPRS subscribers. SGSNs detect new GPRS mobile stations in a given service area, process registration of new mobile subscribers, and keep a record of their location inside a given area. Therefore, the SGSN performs mobility management functions such as mobile subscriber attach/detach and location management. The SGSN is connected to the base station subsystem via a Frame Relay connection to the *packet control unit* (PCU) in the BSC.

GPRS requires modifications to numerous network elements.

GPRS Subscriber Terminals

A totally new subscriber terminal is required to access GPRS services. These new terminals will be backward compatible with GSM for voice calls. New *terminals* (TEs) are required because existing GSM phones do not handle the enhanced air interface, nor do they have the capability to packetize traffic directly. A variety of terminals will exist, as described in a previous section, including a high-speed version of current phones to support high-speed data access, a new kind of *personal digital assistant* (PDA) device with an embedded GSM phone, and PC Cards for laptop computers. All these TEs will be backward compatible with GSM for making voice calls using GSM.

GPRS BSS

A software upgrade is required in the existing Base Transceiver Site (BTS). The Base Station Controller (BSC) will also require a software upgrade, as well as the installation of a new piece of hardware called a packet control unit (PCU). The PCU directs the data traffic to the GPRS network and can be a separate hardware element associated with the BSC. Each BSC will require the installation of one or more PCUs and a software upgrade. The PCU provides a physical and logical data interface out of the Base Station System (BSS) for packet data traffic. The BTS may also require a software upgrade, but typically will not require hardware enhancements. When either voice or data traffic is originated at the subscriber terminal, it is transported over the air interface to the BTS, and from the BTS to the BSC in the same way as a standard GSM call. However, at the output of the BSC, the traffic is separated; voice is sent to the MSC per standard GSM, and data is sent to a new device called the SGSN, via the PCU over a Frame Relay interface.

GPRS Network

In the core network, the existing MSCs are based upon circuit-switched central office technology, and they cannot handle packet traffic. The deployment of GPRS requires the installation of new core network elements called the Serving GPRS Support Node (SGSN) and Gateway GPRS Support Node (GGSN). Figure 3-2 shows some of the overlay elements.

From a high level, GPRS can be thought of as an overlay network onto a second-generation GSM network. This data overlay network provides packet data transport at rates from 9.6 to 171 Kbps. Additionally, multiple

Figure 3-2
The overlay network interworks between public and private networks.

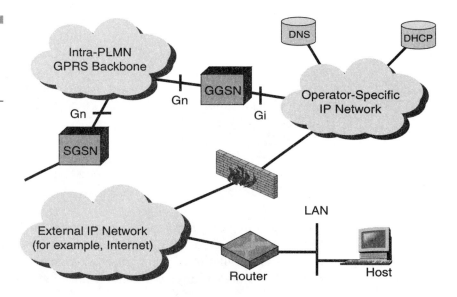

users can share the same air-interface resources. GPRS attempts to reuse the existing GSM network elements as much as possible, but in order to effectively build a packet-based mobile cellular network, some new network elements, interfaces, and protocols that handle packet traffic are required.

Databases (VLR and HLR)

All the databases involved in the network will require software upgrades to handle the new call models and functions introduced by GPRS. The Home Location Register (HLR) and *Visitor Location Register* (VLR) will especially require upgrades to functionally service GPRS because both GSM and GPRS networks must track and monitor the mobile stations. The commonality of the database creates a smoother transition using the central databases to manage and internetwork the two environments. However, the networks have some elements that may not be initially deployed, creating the need to establish a register (database) at the new GPRS serving nodes, such as the SGSN and GGSN. Functionally, the SGSN will act as a VLR. Figure 3-3 provides an overall reference model for the GPRS in a GSM architecture with all associated implementations.

Additionally, enhancements will be made to the *Equipment Identity Register* (EIR) and the *Authentication Center* (AuC) databases to control the security and authentication of the mobile station subscriptions.

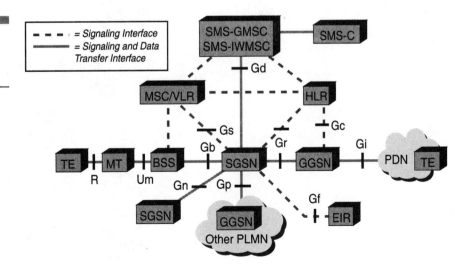

Figure 3-3
The network reference model for GSM.

As mentioned, the *European Telecommunications Standards Institute* (ETSI) achieved a coup when the *American National Standards Institute* (ANSI) in North America also accepted the GPRS model. In the North American model, the past use of the *Telecommunications Industry Association/Electronics Industry Association* (TIA/EIA) standards was the only way that was supported for wireless cellular networks. However, with the ETSI specifications, a new thrust in the international internetworking comes one step closer. The ANSI model for GPRS, although similar and different at the same time, is shown in Figure 3-4.

Data Routing

One of the main issues in the GPRS network is the routing of data packets to/from a mobile user. The issue can be divided into two areas: data packet routing and mobility management.

Data Packet Routing

The main functions of the GGSN involve interaction with the external data network. The GGSN updates the location directory using routing informa-

Figure 3-4
The GPRS reference model for North America (ANSI).

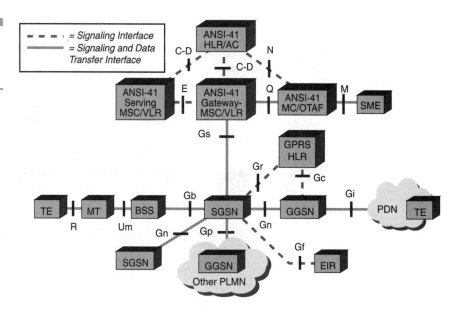

tion supplied by the SGSNs about the location of a *mobile station* (MS) and routes the external data network protocol packet encapsulated over the GPRS backbone to the SGSN currently serving the MS. It also decapsulates and forwards external data network packets to the appropriate data network and collects charging data that is forwarded to a charging gateway. Figure 3-5 shows the use of the various tools in a GPRS network.

Three different routing schemes are possible: mobile-originated message (possibility 1), network-initiated messages when the MS is in its home network (possibility 2), and network-initiated messages when the MS has roamed to another GPRS operator's network (possibility 3). In these examples, the operator's GPRS network consists of multiple GSNs (with a gateway and serving functionality) and an intraoperator backbone network.

GPRS operators will allow roaming through an interoperator backbone network. The GPRS operators connect to the interoperator network via a *Boarder Gateway* (BG), which can provide the necessary interworking and routing protocols (for example, *Border Gateway Protocol* [BGP]). It is also foreseeable that GPRS operators will implement *quality of service* (QoS) mechanisms over the inter-operator network to ensure *service level agreements* (SLAs). The main benefits of the architecture are its flexibility, scalablility, interoperability, and roaming.

The GPRS network encapsulates all data network protocols into its own encapsulation protocol, called the *GPRS Tunneling Protocol* (GTP). This is

Figure 3-5
The various components for data routing.

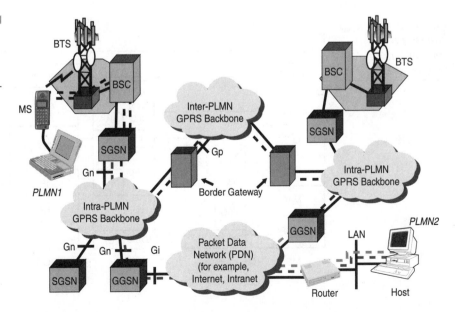

done to ensure security in the backbone network and to simplify the routing mechanism and the delivery of data over the GPRS network.

GPRS Mobility Management

The operation of the GPRS is partly independent of the GSM network. However, some procedures share the network elements with current GSM functions to increase efficiency and to make optimum use of free GSM resources (such as unallocated time slots). Figure 3-6 shows a reference model for the protocols used between the GPRS components. The model is a relay chart for protocols in the *Open Standards Interconnect* (OSI) model.

A mobile station has three states in the GPRS system: idle, standby, and active. The three-state model represents the nature of packet radio relative to the GSM two-state model (idle or active).

Data is transmitted between a mobile station and the GPRS network only when the mobile station is in the active state. In the active state, the SGSN knows the cell location of the mobile station. However, in the standby state, the location of the mobile station is known only as to which routing area it is in. (The routing area can consist of one or more cells within a GSM location area.) When the SGSN sends a packet to a mobile station that is in the standby state, the mobile station must be paged. Because the SGSN

Figure 3-6
The GPRS traffic protocol stack.

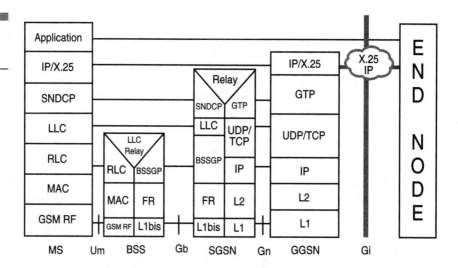

knows the routing area in which the mobile station is located, a packet-paging message is sent to that routing area. After receiving the packet-paging message, the mobile station gives its cell location to the SGSN to establish the active state.

Packet transmission to an active mobile station is initiated by packet paging to notify the mobile station of an incoming data packet. The data transmission proceeds immediately after packet paging through the channel indicated by the paging message. The purpose of the packet-paging message is to simplify the process of receiving packets. The mobile station has to listen to only the packet-paging messages, instead of all the data packets in the downlink channels, reducing battery use significantly.

When a mobile station has a packet to be transmitted, access to the uplink channel is needed. The uplink channel is shared by a number of mobile stations, and its use is allocated by a BSS. The mobile station requests use of the channel in a packet random access message. The transmission of the packet random access message follows Slotted Aloha procedures. The BSS allocates an unused channel to the mobile station and sends a packet access grant message in reply to the packet random access message. The description of the channel (one or multiple time slots) is included in the packet access grant message. The data is transmitted on the reserved channels. The main reasons for the standby state are to reduce the load in the GPRS network caused by cell-based routing update messages and to conserve the mobile station battery. When a mobile station is in the standby state, the SGSN does not need to be informed of every cell change—only of every routing area change. The operator can define the size of the routing area and, in this way, adjust the number of routing update messages.

In the idle state, the mobile station does not have a logical GPRS context activated or any *Packet-Switched Public Data Network* (PSPDN) addresses allocated. In this state, the MS can receive only those multicast messages that can be received by any GPRS mobile station. Because the GPRS network infrastructure does not know the location of the mobile station, it is impossible to send messages to the mobile station from external data networks. A cell-based routing update procedure is invoked when an active mobile station enters a new cell. In this case, the mobile station sends a short message containing information about its move (the message contains the identity of the mobile station and its new location) through GPRS channels to its current SGSN. This procedure is used only when the mobile station is in the active state.

Network Architecture— New Interfaces

The GPRS backbone network will permit point-to-point GPRS calls, inter-working with the BSS, HLR, MSC, SMSC, and the Internet. A new set of interfaces has been developed for GPRS. These interfaces all are labeled with a G_x where the x stands for a variety of interfaces, as discussed in the following section and in Figure 3-7.

These services will be supported via the following interfaces:

- G_b Between the PCUSN and SGSN, using Frame Relay
- G_r Between SGSN and HLR, extension of the *Mobile Application Part* (MAP)
- G_n Between SGSN and GGSN, using the GTP (tunneling) protocol
- G_i Between GGSN and PDNs (X.25 and *Internet Protocol* [IP])
- G_s Between SGSN and MSC/VLR, for some simultaneous GPRS and GSM operations (same as *Base Station Mobile Application Part* [BSSMAP] but optional)
- G_d Delivers SMS messages via GPRS (same as MAP from GSM)
- G_c Between GGSN and HLR (same as MAP but optional)

Figure 3-7
The new interfaces in GPRS.

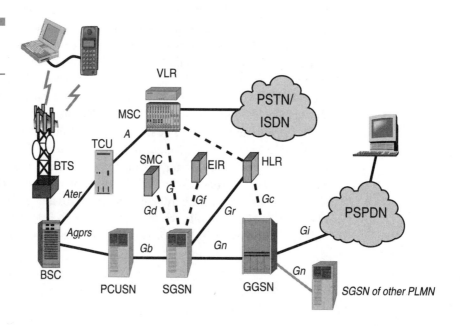

Frames going out from the BTS will be transparently conveyed by the BSC to the PCUSN, which handles GPRS-specific Packet Processing (the normal process uses Frame Relay, but, this may change to *Asynchronous Transfer Mode* [ATM] or other protocols in the future).

The Different Backbones Used

Each SGSN is linked to Packet Control Unit Switching Nodes (PCUSN) with a Frame Relay network, which is

- The only protocol possible (today) in the actual state of the ETSI specifications
- Simpler than X.25
- Capable of supporting data rates up to 2 Mbps

The SGSN and GGSN are linked together within the GPRS backbone based on IP routing. GPRS tunnels the *protocol data unit* (PDU) using the GPRS Tunneling Protocol (GTP). GTP IPv4 is used as a GPRS backbone network layer protocol. Figure 3-8 provides a summary of the various backbone networks used.

The GTP header contains a tunnel endpoint identifier for point-to-point and multicast packets as well as a group identity for point-to-multipoint

Figure 3-8
The different
backbones used.

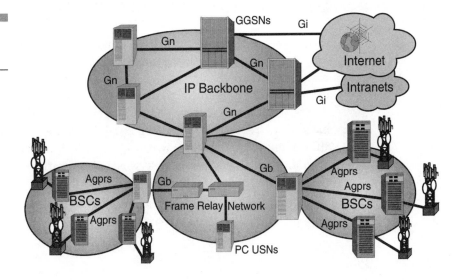

packets. Additionally, a type field in which the PDU type is specified and a QoS parameter is included. Three routing protocols are available:

■ RIP2

■ Static

■ *Open Shortest Path First* (OSPF)

Layer 2 subnetwork architectures that may be used below IP include

■ Ethernet

■ Token Ring

■ *Fiber-Distributed Data Interface* (FDDI)

■ *Integrated Services Digital Network* (ISDN)

■ ATM

GPRS will support interworking of mobile stations with IP first and X.25 later. Further, GPRS will transmit the corresponding PDU transparently by encapsulation and decapsulation.

The G_i interface between *Public Land Mobile Network* (PLMN)/GPRS and the Intranet/*Internet Service Provider* (ISP) is carried out via the public network. *IP Security* (IPSec) protocols may be used to provide authentication and encryption of the link. This enables confidential transport of the G_i interface over the public domains such as the Internet.

Initial Implementations

The first releases of GPRS products must support IP and interworking with the Internet and intranets. Figure 3-9 shows a view of the initial implementation networks. Only one SGSN will be required due to the relatively low number of users in North America.

Interconnection between GGSN and GSM/NSS nodes (MSC/VLR, HLR, and SMSC) requires a *Signaling System Number 7* (SS7)/IP gateway, or SIG, to link the IP backbone with G_s, G_r, and G_d interfaces.

To manage IP addresses, a server that contains the following functions will be used:

■ **Domain Name Server (DNS)**   To translate domain names to IP addresses and vice versa

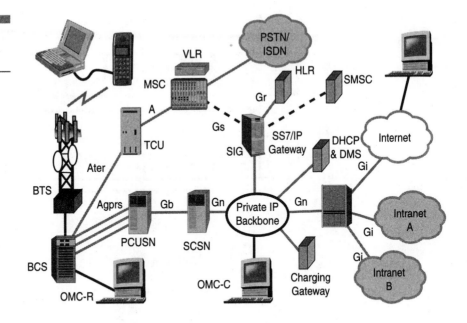

Figure 3-9
Initial
implementations.

■ **Dynamic Host Configuration Protocol (DHCP)** To provide
automatic addressing and readdressing for mobile hosts

TDMA—GPRS Physical Channel Capacity

The *Time Division Multiple Access* (TDMA) frame structure for GPRS is the
same as for GSM and is shown in Figure 3-10. The sequence of all time slots
in a particular position of each TDMA frame is defined to be a physical
channel. A physical channel that has been allocated to GPRS service is
called a *Packet Data Channel* (PDCH). As we shall see later, various combinations of the logical channels can be mapped onto a single physical channel. Physical channels can also be grouped to provide higher data
transmission rates.

GPRS provides for flexible allocation of physical channels to GPRS service. The GPRS traffic load in a given cell varies as a function of time. The
network has the option of dynamically changing the number of physical
channels allocated to GPRS depending on the demand.

Figure 3-10
The physical
channels.

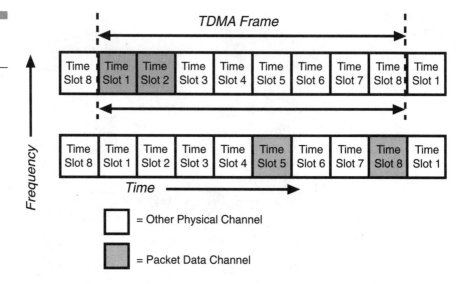

GPRS Logical Channels

In GSM terms, a logical channel refers to a flow of information between entities for a particular purpose. Logical channels are carried within the physical channels (such as PDCHs). The following logical channels are defined and shown in Figure 3-11 for GPRS:

Packet Broadcast Control Channel (PBCCH)

Packet Broadcast Control Channel (PBCCH) is a downlink function used for broadcast of system information to the mobile stations in a cell. If the PBCCH is not used in a cell, then regular *Broadcast Control Channel* (BCCH) may be used to send packet data-specific broadcast information.

Packet Common Control Channel (PCCCH)

Packet Common Control Channel (PCCCH) is a common control channel service that is comprised of the following logical channels for common channel signaling for the packet data:

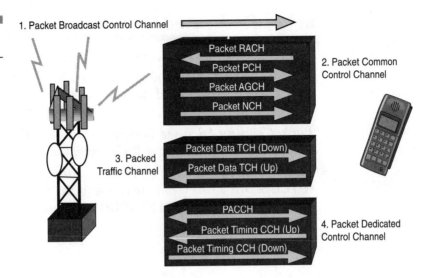

Figure 3-11
The logical channels

- *Packet Random Access Channel* **(PRACH)** An uplink function used by the mobile station to access the system and send control information of traffic packet data.
- *Packet Paging Channel* **(PPCH)** A downlink function used for paging the mobile station for mobile-terminated communications. The PPCH may be shared for packet data as well as circuit data services to page the mobile station.
- *Packet Access Grant Channel* **(PAGCH)** A downlink function used to assign radio resources to the mobile station during call setup.
- *Packet Notification Channel* **(PNCH)** A downlink function used to send a notification of the point-to-multipoint multicast to a group of mobile stations prior to sending the data. The point-to-multipoint service is not specified in Phase I of GPRS, but was addressed as part of Phase II.

Packet Data Traffic Channel (PDTCH)

The traffic channel is an up and downlink function used for user data traffic transfer. The PDTCH is temporarily dedicated to a user or group of users (for multipoint). PDTCH for uplink and PDTCH for downlink are unidirectional and assigned separately to support asymmetric user traffic flow.

Packet-Dedicated Control Channel (PDCCH)

- ***Packet Associated Control Channel* (PACCH)**   This is an uplink and downlink function used to carry signaling information to and from the mobile station. PACCH shares resources with the PDTCH assigned to the mobile station.

- ***Packet Timing Advance Control Channel/Uplink* (PTCCH/UL)** This is used for estimation of timing advance of one mobile station.

- ***Packet Timing Advance Control Channel/Downlink* (PTCCH/DL)**   This is used to transmit timing advance information to several mobile stations. One PTCCH/DL is paired with several PTTCH/ULs.

Mapping Logical Channels onto Physical Channels

As stated earlier, logical channels are carried on physical channels. In fact, multiple logical channels can be mapped onto the same physical channel in a timesharing fashion using a superframe structure. Several combinations of logical channels can be multiplexed onto the same physical channel; these are shown in Figure 3-12. Three possible combinations are allowable:

Broadcast Control + Common Control + Traffic + Dedicated Control
            PBCCH + PCCCH + PDTCH + PACCH + PTCCH

Common Control + Traffic + Dedicated Control
            PCCCH + PPDTCH + PACCH + PTCCH

Traffic + Dedicated Common Control
            PCTCH + PACCH + PTCCH

Note that the Packet Common Control Channel (PCCCH) is made up of the following logical channels:

            PCCH = PAGCH + PPCH + PRACH + PNCH

Figure 3-12
Mapping the logical channels on the physical channels.

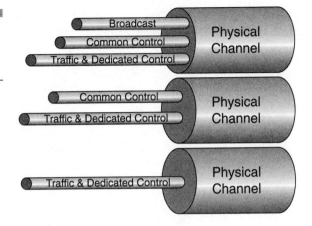

The combinations previously listed are not dissimilar to the GSM architecture, although they deal with the packet data channels and control functions. One can see several similarities between the GSM and GPRS architectures. As stated earlier, GPRS is merely an overlay of GSM so the channel definitions are close.

Function of GPRS Elements

Objectives

When you complete the reading in this chapter, you will be able to

- Describe the locations for the CCU.
- Describe the four different types of channel codec services.
- Understand where the options are for locating the PCU.
- Discuss the role of the GGSN and SGSN.
- Understand the way that the HLR and VLR interact in the GPRS network.

For the first time, *General Packet Radio Service* (GPRS) fully enables mobile Internet functionality by permitting interworking between the existing Internet and the new GPRS network. Any service that is used over the fixed Internet today (such as *File Transfer Protocol* [FTP], Web browsing, chat, e-mail, and telnet) will be as available over the mobile network because of GPRS. In fact, many network operators are considering the opportunity to use GPRS to help become *Wireless Internet Service Providers* (W-ISPs) in their own right.

The World Wide Web is becoming the primary communications interface —people access the Internet for entertainment and information collection, the intranet for accessing company information and connecting with colleagues, and the extranet for accessing customers and suppliers. These are all derivatives of the World Wide Web aimed at connecting different communities of interest. The trend is moving away from storing information locally in specific software packages on PCs to remotely on the Internet. When you want to check your schedule or contacts, instead of using a dedicated contact manager on the PC, you go onto the Internet site such as a portal. Hence, Web browsing is a very important application for GPRS. Because it uses the same protocols, the GPRS network can be viewed as a subnetwork of the Internet with GPRS-capable mobile phones being viewed as mobile hosts. This means that each GPRS terminal can potentially have its own *Internet Protocol* (IP) address and will be addressable as such. This really depends on the implementation by the operator. The impact to the overall *Global Systems for Mobile* (GSM) operation will dictate what and how much influence the operator applies to the GPRS network initially.

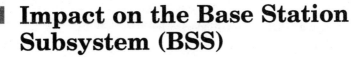

Impact on the Base Station Subsystem (BSS)

No major changes occur inside the *Base Station Subsystem* (BSS). However, two functions must be inserted inside the BSS for the GPRS functionality to work. These are listed in the following section and shown in Figure 4-1:

- *Channel Codec Unit* (CCU)   Deals mainly with the new coding schemes (as well as any compression coding techniques)
- *Packet Control Unit* (PCU)   Responsible for providing the interface with the GPRS network (Frame Relay) and managing time slot allocation

The CCU is located in the *Base Transceiver System* (BTS) without any hardware impact. The PCU can be located at the *Network Service Subsystem* (NSS) side or at the *Base Station Controller* (BSC) side, depending on the network engineering rules or any cost constraints (such as long-distance *Pulse Coded Modulation* [PCM] interfaces). However, it is a part of the BSS.

The *European Telecommunications Standards Institute* (ETSI) standard for GPRS, GSM 03.60, recommends three positions for the PCU. In many

Figure 4-1
Different places to locate the CCU and PCU.

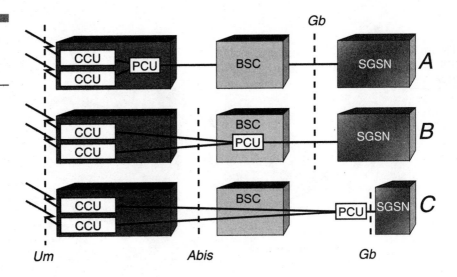

implementations, the PCU is located in the *Packet Control Unit Support Node* (PCUSN), which resides between the BSC and the *Serving GPRS Support Node* (SGSN). Thus, the PCUSN is aligned with the second and third options. The advantage of placing the PCU out at the CCUs is that the round-trip delay is shorter. The advantage of having the PCU located at the SGSN is that one PCUSN can provide service and control for as many as 12 BSCs. Each BSC can control up to 128 BTS systems (although installations typically are less than that).

The Packet Control Unit Support Node (PCUSN)

The PCUSN (Figure 4-2) is a new functional unit defined as part of the GPRS specifications. The PCUSN is a stand-alone node in the BSS. Its main function is to complement the BSCs (both 2G and 3G wireless standards) with the PCU capability.

The PCUSN is typically connected to the BSC with the proprietary A_{gprs} interface and to the SGSN via the standard G_b interface. Its main function is to provide the interworking function between the two interfaces. The PCUSN can be connected to a single BSC or multiple BSCs. However, all

Figure 4-2
The PCUSN is a
new device.

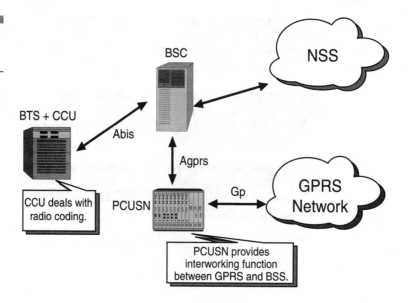

signaling and traffic channels from the BSC must pass through the same PCUSN.

Channel Codec Unit (CCU)

The role of the CCU is to perform the following functions:

■ The channel-coding functions (such as forward error correction and interleaving on the air interface)

■ Radio channel measurements (quality of receive signal)

■ Radio management

The CCU is located in the BTS, in a choice of software-only upgrades or with hardware and software upgrades depending on the coding being performed. The initial releases of the CCUs support only a limited number of coding schemes. The four choices are numbered CS-1 through CS-4. These are shown in the graph in Figure 4-3 and described in the following list:

■ CS-1 offers a data rate between 8 Kbps to 64 Kbps.

■ CS-2 offers a data rate between 12 Kbps and 96 Kbps.

■ CS-3 offers a data rate between 14.4 Kbps and 96+ Kbps.

■ CS-4 offers a data rate between 20 Kbps and 115+ Kbps.

Figure 4-3
The four coding rates and data speeds expected.

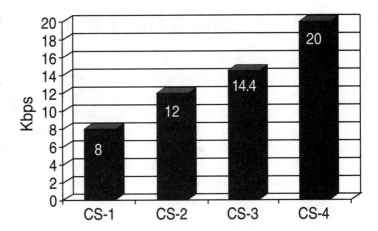

This all depends on the *mobile station* (MS) equipment capabilities and the number of time slots accepted (multiclass). Higher data rates are far more sensitive to radio link quality:

- CS-1 is mandatory for the BSS and is also used for signaling.
- CS-1, CS-2, CS-3, and CS-4 are mandatory for mobile stations.
- CS-4 has no forward error correction.

Data Link Layer—Layer 2

In GSM, only Layers 1 and 2 are directly related to the air interface where the *Link Access Protocol on Dm channel* (LAPDm) signaling is used between the mobile station and the BTS. Other interfaces within a GSM network use *Link Access Procedure on the D channel* (LAPD) and *Signaling Systems Number 7* (SS7). The link layer uses LAPDm, a modified version of LAPD used in *Integrated Services Digital Network* (ISDN). In general terms, the link layer receives services from the physical layer and provides services to the network layer. The services to Layer 3 are provided via a *Service Access Point* (SAP), where each point is given a separate identifier called the *Service Access Point Identifier* (SAPI). One or more endpoints can be associated with a SAP, which are identified by a *Data Link Connection Identifier* (DLCI). Peer-to-peer protocols exist between the mobile station and the BTS and the virtual connection between the two is called the Data Link Connection.

LAPD Data Link Layer

At the Layer 2 protocol, we use the 184 bits of user information. This is input from the application or higher-level protocols in the stack. The 23 octets are received and channel coded for the link using a convolutional coding technique, which results in the generation of 57 octets. Now the 57 octets (456 bits) are truncated into the transmission unit of twice the 57 bits plus overhead. Four sets of transmission are sent in the time slots in the four *Time Division Multiple Access* (TDMA) time slots. This Layer 2 protocol prepares the data for the link so that it can be properly transmitted and protected.

Impact on BSC: A New LAPD

The A_{gprs} interface is composed of a dedicated A_{gprs} PCM (more than one can be used in the future) 64-Kbps channel. It carries

- One signaling 64-Kbps LAPD channel (TS-24 if used on a T1 or TS-16 or 31 if used on an E1), which conveys three different SAPIs:

  - A_{gprs} OML for *operation and maintenance links* between the BSC and PCU
  - A_{gprs} GSL for *GPRS radio signaling links* between the BSC and PCU
  - A_{gprs} RSL for *radio signaling links* between the BSC and PCU

Figure 4-4 shows the functional view of the LAPD link. The Link Access Procedure for Data channel is a function of ISDN that is modified for GPRS.

Figure 4-4
The new LAPD
for GPRS.

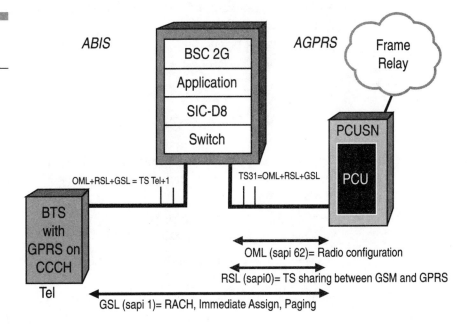

A$_{gprs}$ OML BSC-PCU

The BSC and PCU terminate the OML. It conveys all messages dedicated to *operation and maintenance* (OAM) for radio-related issues. It is primarily composed of

- Cell and TDMA configuration from the BSC to PCU to indicate *operational and maintenance control for radio* (OMC-R) configuration regarding radio-related issues (cell properties or number if static time slot is used for GPRS)
- Mapping of static time slots to A$_{gprs}$ interface from the PCU to the BSC

A$_{gprs}$ GSL BTS-PCU (Through the BSC)

The A$_{gprs}$ GSL is concentrated by the BSC. The BTS terminates the U$_m$ *Common Control Channels* (CCCHs) and forwards all the GPRS-specific messages on the GSL to the BSC that concentrates all BTS messages to the PCU. It is mainly composed of

- The channel request from the mobile station for *Temporary Block Flow* (TBF) Establishment Granted
- Paging from PCU to the mobile station for TBF Downlink Establishment Request

A$_{gprs}$ RSL BSC-PCU

The A$_{gprs}$ RSL conveys all the messages dedicated to the allocation of GPRS time slots and dynamic radio time slots between GSM and GPRS. It is mainly composed of

- Indication from BSC to PCU to define the availability/unavailability of radio TDMA and radio cell for GPRS
- Time slot requests from the PCU to the BSC to get more GPRS resources
- Time slot grant/recover from the BSC to the PCU to give back or get GPRS resources for GSM purposes

■ Conveys all the messages sent by the mobile station over the *stand-alone dedicated control channel* (SDCCH) for GPRS purposes (addressing GSM/GPRS capabilities for class B mobile stations)

■ GPRS suspend/resume from the BSC to the PCU in order to activate/deactivate service in GSM transfer

Function of the PCUSN

The main role of the PCU, as shown in the architecture in Figure 4-5, is to provide the interworking function between the two interfaces as follows:

■ The packetized radio interface A_{gprs} (synchronous connection-oriented link)

■ The packet network interface G_b (asynchronous and connectionless)

The PCU is responsible for the GPRS RLC/MAC layer function and

■ GSM radio frequency

■ L1bis

Figure 4-5
The PCUSN.

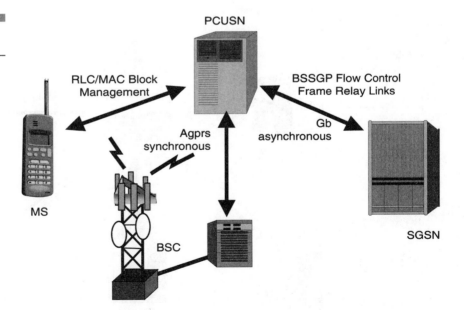

- Network services
- *BSS GPRS Protocol* (BSSGP)
- Relay function

The PCUSN is introduced as a separate node in the BSS in order to provide the PCU functionality.

Serving GPRS Support Node (SGSN) Functions

The SGSN, serving as the packet version of the *Mobile Switching Center/Visitor Location Register* (MSC/VLR), requires the services of a packet switch in order to properly perform its role in the GPRS network. Consequently, the SGSN is hosted in a high-end switching system. Figure 4-6 shows the relationship of the SGSN in a GPRS network and the functions performed. The SGSN performs the following functions:

- *Mobility management* (MM)
- Routing of packet data using the *Internet Protocol* (IP) at Layer 3

Figure 4-6
The Serving GPRS
Support Node
(SGSN).

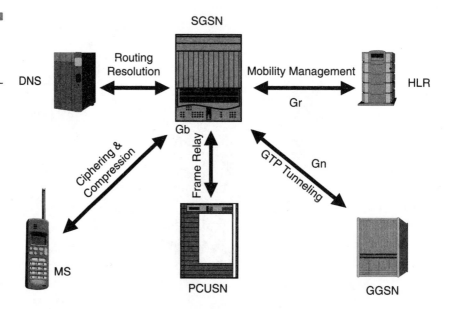

- Authentication
- Encryption
- Compression

MM performs the following tasks as a part of the overall scheme:

- Session management
- State control; mobile station state
- Data packet routing on the downlink, including location tracking

The SGSN performs authentication and cipher-setting procedures based on the same algorithms, keys, and criteria as in existing GSM; however, the ciphering algorithm is optimized for packet data transmission.

Additional functions of the SGSN, shown in Figure 4-7, include the following:

- Temporary storage capability

  - Tracks the mobile station location by routing areas or cell location
  - Keeps track of the connected *Gateway GPRS Support Node* (GGSN) so that it can determine what GGSNs can be used for connections to the IP networks, intranets, the Internet, or X.25 networks
  - Maintains a log of the active *Packet Data Protocol* (PDP) contexts (this is a VLR functional equivalent)

Figure 4-7
The additional functions of the SGSN.

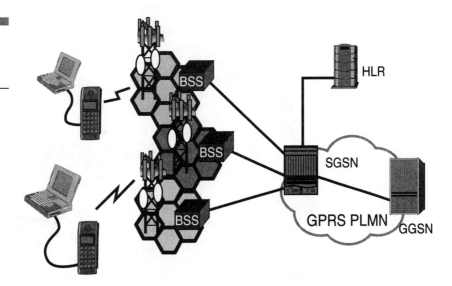

- SGPRS attach

  - Receives the MS's subscription info from the *Home Location Register* (HLR)
  - Notifies the old SGSN of MS attach request
  - Verifies response to identity request

- Initiates a detach request; notifies MSC/VLR of detach procedure
- Authentication
- User's subscription info (*International Mobile Subscriber Identity* [IMSI] or SGSN # attached to)

Gateway GPRS Support Node (GGSN)

The GGSN (Figure 4-8) serves as the interconnect point between the SGSN and the external packet data network, requiring features to provide secure communications between GPRS users and IP. Moreover, the GGSN provides tunneling capabilities within the GPRS network system itself. The

Figure 4-8
The GGSN.

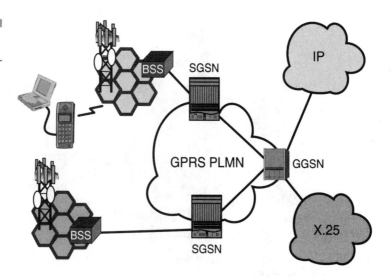

GGSN also performs a function similar to that of the *Gateway MSC* (GMSC) and is very close to that of a router in IP terms.

Within the GPRS backbone, IP and X.25 data packets are encapsulated in the *GPRS Tunneling Protocol* (GTP):

■ From a mobile station to the external data network, the GGSN strips off the GTP and lower-layer headers. Then, it delivers the IP and X.25 packet in its native form to the external network nodes.

■ From the external data network to a mobile station, the GGSN performs the opposite operation: it adds GTP and lower-layer headers, followed by transporting the packet to the appropriate SGSN.

Tunneling is the transfer of encapsulated data units within the *Public Land Mobile Network* (PLMN) from the point of encapsulation to the point of decapsulation. A tunnel is a two-way point-to-point path; only the endpoints are identified.

Home Location Register (HLR)

The HLR is an existing GSM network element. It is responsible for keeping track of the mobile's location in the network (such as corresponding SGSN) as well as tracking the activity status (active or inactive) for mobiles in its domain. In order to support GPRS services, the HLR must be enhanced to include GPRS subscription data, including subscribed *quality of service* (QoS), statically allocated PDP addresses, and roaming permissions.

The HLR is the main network database. Only one HLR is present (logically) in any network, although it may be distributed. The information stored relates to all subscribers registered in the network. The presence of the information is independent of the location of the subscriber. Information contained in the HLR includes the Subscriber ID, MSISDN, IMSI, current location (switch area), and subscription services chosen. The HLR is connected to the MSC across the C interface. The HLR also plays a role in the mobile station attach function. When a mobile station attempts to attach to the GPRS network, the HLR provides subscription data to the SGSN. The HLR is updated to maintain information regarding the user's subscription services for GPRS. Figure 4-9 shows the HLR.

Figure 4-9
The HLR.

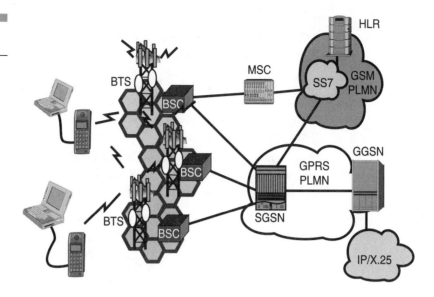

The Visitor Location Register (VLR)

The VLR holds similar information as the HLR. However, the VLR is always distributed (that is, one is associated with each MSC). The information contained in this database is temporary and will only be available as long as the subscriber is in the area. The information relates to all subscribers in the MSC area only; anyone else in another MSC area is dropped from the database. The VLR also contains details of foreign mobile subscribers roaming in its area.

In addition to the information contained in the HLR information, the VLR also contains additional information:

■ Stored authentication parameters

■ Location area identity

■ Routing area identity

■ *Mobile Station Roaming Number* (MSRN)

■ *Temporary Mobile Subscriber Identity* (TMSI) and *Packet TMSI* (P_TMSI)

The VLR is connected to the MSC using a B interface, as shown in Figure 4-10.

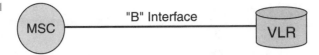

Figure 4-10
The function of
the VLR.

"B" Interface

MSC — VLR

⭐ Holds similar information as HLR
⭐ One-to-one relationship per MSC
⭐ Temporary storage
- LAI
- Stored Authentication Information
- Mobile Station Roam #
- TMSI or P_TMSI

Other Network Elements

The intra-PLMN backbone is a private network for GPRS users, as shown in Figure 4-11. It connects several SGSNs and GGSNs together within the same GPRS PLMN. The inter-PLMN network is capable of connecting multiple intra-PLMN backbone networks. Border Gateways provide an interface between the inter-PLMN and intra-PLMN backbone. The Border Gateway enhances the security of the network. The Border Gateway concept is beyond the overall scope of GPRS. However, it may be used to uphold roaming agreements between different networks.

Figure 4-11
Other network
elements.

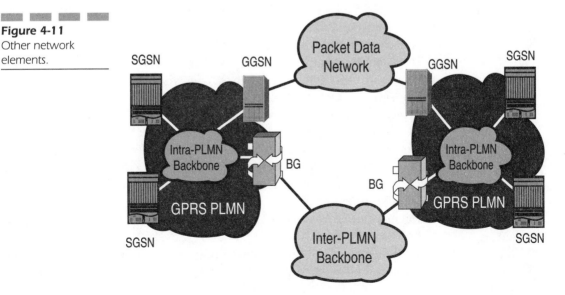

SS7/IP Gateway Functions

The MSC is the central-switching function of the GSM network. The MSC is connected to a SS7 network for the purpose of signaling and performing database queries. The SS7 network uses a network node called the *Signal Transfer Point* (STP), which is a packet-switching node (can be SS7, IP, or X.25). Using a 64-Kbps channel connection between STPs, the network can process its signaling information. Next in a SS7 network is the use of the *Signal Control Point* (SCP) that houses the databases congruent to the network. In many cases, these databases interact with the HLR, VLR, *Equipment Identity Register* (EIR), *Authentication Center* (AuC), and *Public-Switched Telephone Network* (PSTN) nodes. The SCP is used whenever a Global Title Translation is required, which converts numbers ([800] 322-2202 equates to [480] 706-0912) and whenever the *Mobile Application Part* (MAP) is used. These services link across an SS7 interface.

In order to communicate to the legacy-based, circuit-switched nodes (MSC, HLR, SMSC, and so on), according to GPRS standards, SGSNs are supposed to use MAP/BSSAP+ on SS7. Because of the emphasis on the Internet and IP protocols, several providers have chosen to develop MAP/BSSAP+ over IP in the SGSN using a separate SS7 Gateway server to perform the conversion between MAP on IP to MAP on SS7 *Message Transfer Part* (MTP). The SS7/IP Gateway server, or (SIG), as shown in Figure 4-12, provides interworking between GPRS nodes in an IP network and GSM nodes in an SS7 network. Because multiple SGSNs exist in the GPRS network, the SIG is responsible for routing messages from the GSM HLR to the correct SGSN.

Additionally, the SIG converts *Transaction Capabilities Application Part / Global Systems for Mobile* (TCAP/GSM) MAP-encoded messages originating from the GSM HLR in the SS7 network to *User Datagram Protocol / Internet Protocol* (UDP/IP) messages containing the GSM MAP client interface for GSM messages destined for the GPRS SGSN nodes. It performs the reverse messaging that is originated by the SGSN and destined for the HLR. This interface is called the G_r. The SS7 Gateway resides on a high-availability duplex server and is scalable from one to four processors. Each processor is planned to support the traffic load for 250,000 to 300,000 GPRS subscribers.

The normal SS7 network uses the bottom three layers in what is called the *Message Transfer Part 1-3* (MTP1-3). These parts use a different layer of the OSI model to provide the routing and data-link layers across the physical link. Between Layer 3 and the applications layer is the *Signaling*

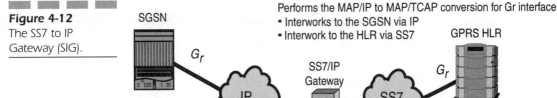

Figure 4-12
The SS7 to IP
Gateway (SIG).

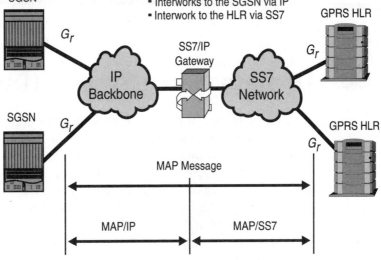

Connection Control Part (SCCP) that is used when database queries are required and when providing both connection and connectionless access to the SS7 networks. The combination of the MTP1-3 and SCCP creates what is called the actual Message Transfer Part.

When looking at the previous layers, the SS7 protocols support the use of the following:

- *Telephony User Part* (TUP) is used for a voice circuit-switched call across the PSTN.

- *ISDN User Part* (IUP) is a newer implementation and replaces the TUP.

- TCAP is an application layer that supports the features and functions of a network.

- MAP sits on top of the TCAP as a means of supporting the different application service entities for mobile users.

- *Base Station Systems Application Part* (BSSAP) is a combination of the BSSMAP and the *Direct Transfer Application Part* (DTAP).

- *Base Station Systems Mobile Application Part* (BSSMAP) transmits messages that the BSC must process. This applies generally to all messages to and from the MSC where the MSC participates in *radio resource* (RR) management.

- DTAP transports messages between the mobile and the MSC, where the BSC is just a relay function, transparent for the messages. These messages deal with MM and *connection management* (CM).

Domain Name System (DNS)

A GPRS network will likely be connected to the Internet. Each registered user who wants to exchange data packets with the IP network gets an IP address. The IP address is taken from the address space of the GPRS operator. In order to support a large number of mobile users, it is essential to use dynamic IP address allocation (in IPv4). Thus, a *Dynamic Host Configuration Protocol* (DHCP) server is installed. The address resolution between IP address and GSM address is performed by the GGSN, using the appropriate PDP context. The routing of IP packets are tunneled through the intra-PLMN backbone with the GTP. Moreover, a *Domain Name Server* (DNS) managed by the GPRS operator or the external IP network operator can be used to map between external IP addresses and hostnames. To protect the PLMN from unauthorized access, a firewall is installed between the private GPRS network and the external IP network. With this configuration, GPRS can be seen as a wireless extension of the Internet all the way to a mobile station or mobile computer. The mobile user has a direct connection to the Internet.

To exchange data packets with external *packet data networks* (PDNs) after a successful GPRS attach, a mobile station must apply for one or more addresses used in the PDN: for example, for an IP address in case the PDN is an IP network. This address is called the PDP address. For each session, a so-called PDP context is created, which describes the characteristics of the session. It contains the PDP type (for example, IPv4), the PDP address assigned to the mobile station (10.0.0.1), the requested QoS, and the address of a GGSN that serves as the access point to the PDN. This context is stored in the MS, the SGSN, and the GGSN. With an active PDP context, the mobile station is visible for the external PDN and is able to send and receive data packets. The mapping between the two addresses, PDP and IMSI, enables the GGSN to transfer data packets between PDN and MS. A user may have several simultaneous PDP contexts active at a given time. The allocation of the PDP address can be static or dynamic. In the first case, the network operator of the user's home-PLMN permanently assigns a PDP address to the user. In the second case, a PDP address is assigned to the user upon activation of a PDP context. The PDP address can be assigned by the operator of the user's home-PLMN (dynamic home-PLMN PDP address) or by the operator of the visited network (dynamic visited-PLMN PDP address). The home network operator decides which of the possible alternatives may be used. In case of dynamic PDP address assignment, the GGSN is responsible for the allocation and the activation/deactivation of the PDP addresses. It will use a combination of DNS and DHCP protocols to facilitate these interactions.

The *Domain Name System* (DNS) and DHCP server is a multiadministrator database application for IP addressing, DNS, and DHCP management, as shown in Figure 4-13. It eliminates problems typically associated with IP management by automating and integrating address assignment.

- You do not need to manually configure addresses and no errors appear.
- No duplication of address assignment occurs.
- Dynamic DNS updates are performed by the server.

The Domain Name Server is a distributed Internet/intranet directory service that translates domain names to IP addresses and vice versa. The lists of domain names are distributed over the Internet in a hierarchy (tree structure) of authorities (name servers).

The Domain Name System is the addressing system of the Internet. Using DNS, your computer determines what IP address (for example, the fictional address 192.168.10.2) corresponds with a particular computer

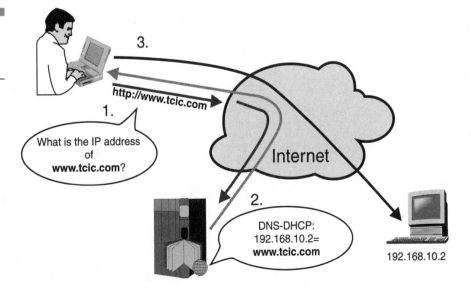

Figure 4-13
The DNS and DHCP functions in GPRS networks.

hostname (for example, **www.tcic.com**). Your computer learns how to get to any IP address on the Internet, and uses that IP address to determine where it should send messages. The Domain Name Server is responsible for maintaining the addresses of all networks and nodes and the IP address translations for them. This server is a distributed name/address mechanism used on the Internet.

A domain is part of the Internet naming hierarchy. Syntactically, an Internet domain name consists of a sequence of names (labels) separated by periods: for example, **mail.smtp.idsweb1.com.us**.

When a user enters a plain-text address, the Domain Name Server is called upon to translate the text-based address into an IP addressing scheme. The DNS then returns the numerical IP address using the address.subnet.node using the dot delimiter to define the user's actual IP address.

The DNS servers are spread around the country and are using a fully distributed computing architecture to maintain addressing information of all registered domains. The databases can be updated on a 30-minute timer, or other time limit as set by the ISP. Normal operation requires at least two DNS connections for redundancy purposes. Regional ISPs provide local DNS servers updated from the master database maintained by Network Solutions, Inc. of Virginia. The InterNic (the name assigned to Network Solutions as the custodian of the numbering plan) assigns names and addresses, then promulgates the changes on a regular basis, keeping all the databases as current as possible. Separate DNS systems are used for the extensions of the addresses. Six major extensions are available:

- **.com** For commercial organizations
- **.gov** For government bodies
- **.edu** For educational institutions
- **.mil** For military organizations
- **.org** For organizations that do not fit the .com role, usually nonprofit organizations
- **.net** For systems performing network services

A domain is also assigned for each country such as Canada (.ca). An organization may fit into more than one category and can choose whichever naming domain it prefers. Now newer roots in the Internet are being created like .store and so on.

We could use the IP address instead of the domain name in some cases. For example, the URL 192.168.10.2 will take you to the **tcic.com** front page, and e-mail sent to bud@[192.168.10.2] will reach **bud@tcic.com**.

(The brackets in the e-mail address are necessary when using IP addresses.) Mapping between domain names and IP addresses using DNS makes things much easier to understand though. They also help with portability; you don't need to retain control over a particular part of a network to maintain your e-mail address or Web services—people can follow you around using your domain name.

Figuring Out Which Server Knows What

When you try to connect to a Web page, for example, **http://www.tcic.com/**, your Web browser splits the URL into its component parts, and determines which part (in this case, **www.tcic.com**) is the hostname. (We'll refer to the host you are trying to reach as the target.) If you've visited a page on the target recently, your computer will remember the IP address of the host, and send a request for the page to that IP address.

If your computer doesn't have the target's IP address, it will connect to whatever local *nameserver* (NS) you have configured it to use. (Actually, it will most likely connect to one of several local nameservers you've configured.) The local nameserver most likely serves multiple machines in your (network) area. If any of the machines served by this DNS server have asked for the target machine recently, the local nameserver will have that machine's IP address stored, and will immediately return that IP address to your computer. If the server does not have the IP address stored, it will try to figure out what remote nameserver has information on the target computer, and retrieve information from there.

The first place your local nameserver will ask for information will be one of the root nameservers—1 of 13 computers that stands at the center of the Domain Name System. Every nameserver on the Internet (with limited exceptions) has the IP addresses of these root servers permanently stored. The root nameservers contain information on which nameservers are responsible for which Internet top-level domains (.com, .org, .gov, .edu, and so on). If you're looking for **www.tcic.com**, the root server that your local nameserver contacts will point you to several nameservers that contain authoritative information for the .com top-level domain.

Once it has the .com server's IP address, your nameserver will ask it for the IP address of the nameserver that has authority over the **tcic.com** domain. The IP address that is returned at this point will be one of the addresses that the domain owner entered when registering the domain with Network Solutions or one of the other registrars.

Now that your local nameserver knows where to find the nameserver for the target machine, it asks that nameserver for the IP address of the target. The target's nameserver returns that information, as well as a *time-to-live* (TTL)—the amount of time that your local nameserver should store the IP address it has received. (This time is generally set fairly low—a matter of days or hours.) Once your local nameserver has this information, it returns the information to your computer, and you are able to connect to the target machine.

DNS uses client/server architecture to maintain and distribute host-names and IP addresses on networks ranging from small *Local Area Networks* (LANs) to the entire Internet. Under DNS, the Internet consists of a hierarchy of domains. This hierarchy, referred to as the *domain name space*, is organized as an inverted tree radiating from a single root, much like a UNIX file system.

Domain Name Space

The root domain (.) is the base of the tree. Final attempts to resolve names to IP addresses take place here if lower-level servers do not have the requested data. The root domain is usually omitted from domain names. Usually, this doesn't affect looking up IP addresses; however, the period (.) is usually vital when configuring DNS data.

Domains Internet uses naming convention called domain names. The domain name consists of two or more parts separated with a period (.). It starts from the least significant domain and ends up to most significant domain or top-level domain. This naming convention naturally defines a hierarchy.

Zones Domain Name Service (DNS) is a huge distributed database that contains information of each domain name. Each server maintains a part of the database called zone. Usually, a zone contains information of one domain. However, one zone may contain information about many (sub)domains.

Each information element is stored in a record that contains at least a domain name and type and type-specific information.

Delegation When a part of a zone is maintained separately, it is delegated to a new nameserver that will have authority of that part of domain name space. The original zone will have a nameserver record for the delegated domain and the new subzone will have a new *Source of Authority* (SOA) record.

Client   DNS client is implemented as a resolver library. Application programs use function calls like *gethostbyname* to find an IP address representing a domain name. The name may be specified only partially and in that case, the resolver library appends a configured local domain name(s) at the end of the name. For example, the user may give the following command:

> *ping hobbes*

The resolver library appends the domain search list and will queries the nameserver with the following:

> *'hobbes.sonera.com hobbes.sonera.fi hobbes'*

Domain names ending with a period are called fully qualified domain names. Search list components are not appended on these names.

Server   DNS server takes care of name service queries sent by clients. The query is answered by using either locally stored information or by asking the information from other nameservers. Sending queries to other nameservers is potentially time and network resources consuming task. Storing previously queried information in a local cache optimises the process. Each nameserver record has a TTL that specifies the time they may be cached. When TTL expires, the record is discarded and a new query is performed.

Servers build a hierarchy. At the top of the hierarchy are root nameservers. They have information about all top-level domain nameservers like .net or .fi nameservers. These nameservers, in turn, know about all nameservers immediately under their domain.

One nameserver can serve several domains. Several nameservers may also serve one domain. In fact, at least two nameservers for each domain are strongly recommended. This ensures service for the domain in case one of the nameservers is temporarily out of order. One of the nameservers serving a domain contains the master or primary copy of the zone information. All changes are made to this copy. Other nameservers are slave or secondary nameservers for this domain.

In GRPS, the internetwork operations between two or more PLMNs are extremely important to handle the demands of a roaming user. Therefore, DNS is a crucial service and has a significant impact on the operation of the network.

DNS and Inter-PLMN Network

Each PLMN operator should have at least two DNS servers. This makes it possible to upgrade one of the servers without service interruption. The servers should keep a cache of recently queried DNS records. Caching reduces query response time and decreases network traffic. A necessary DNS hierarchy can be arranged through two possibilities. The first is to configure the nameserver of each domain at the inter-PLMN network individually at each PLMN operator. Each time a new domain is added to the inter-PLMN backbone network name service or any authoritative nameserver address is changed every operator must update the DNS servers. Over time this will become tedious and can become the likely source for operational roaming problems.

Another alternative is to have common a GPRS root nameserver. Every change in domain or DNS information is updated at the master GPRS root nameserver and the changed information is immediately active. Because the GPRS root nameserver is critical for operation, it will normally be replicated at several locations in the inter-PLMN backbone network as a matter of prudent operation. GPRS root nameservers should contain necessary information to reach the individual operator DNS servers. Root server security is crucial. For example, they may only provide zone transfer to other GPRS root nameservers.

Dynamic Host Configuration Protocol (DHCP)

DHCP is also based on a client-server architecture, whereby the DHCP client (such as a desktop computer) contacts a DHCP server for configuration parameters. The DHCP server is typically centrally located and operated by the network administrator. Because a network administrator runs the server, DHCP clients can be configured reliably and dynamically with parameters appropriate to the current network architecture.

The most important configuration parameter carried by DHCP is the IP address. A computer must be initially assigned a specific IP address that is appropriate to the network to which the computer is attached and that is not assigned to any other computer on that network. If a computer moves to a new network (such as a GPRS or other mobile network), it must be assigned a new IP address for that new network. DHCP can be used to automatically manage these assignments.

DHCP carries other important configuration parameters such as the subnet mask, default router (the GGSN address), and Domain Name System (DNS) server. Using DHCP, a network administrator can avoid hands-on configuration of individual computers through complex and confusing setup applications. Instead, those computers can obtain all required configuration parameters automatically, without manual intervention, from a centrally managed DHCP server. The DHCP breaks the human/machine association by providing automatic addressing/readdressing for mobile hosts, as shown in Figure 4-14. When a client joins the network, it sends a request for an IP address to the DHCP server, which assigns host configuration information to this client from a pool of dynamic addresses. By dynamically assigning the IP address and configuration information to networked devices, DHCP reduces the administrative burden of manually configuring computers for network use.

How the Protocol Works (Basic)

In its simplest form, the client sends a request for a server (optionally, with its suggested IP address). The server responds with an available IP address. Next, the client sends a request to the selected server for its configuration options. Finally, the server responds with the client's committed

Figure 4-14
DHCP in action.

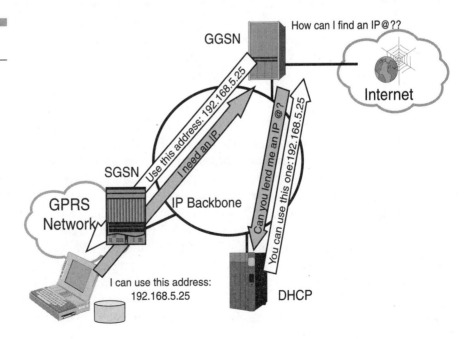

IP address along with other options such as its net mask. If a router exists between the client and the server, the router should use a BOOTP forwarding agent to get the request from the client to the server and back.

Computer networks always seem to be changing. New devices (PCs, printers, and so on) must be attached, old devices disconnected, new branches added, mobile workers hosted, and temporary employees accommodated; all of these changes can happen every day. Managing all that change can prove a major undertaking without systems that respond automatically to changing demands. On a *Transmission Control Protocol / Internet Protocol* (TCP/IP) network, each system must have an IP address, subnet mask, and router address (at a minimum) in order to communicate.

Charging Gateway Function

Make no mistake about it: The charging function is a crucial component of a GPRS network. Earlier the discussion led to the fact that the various operators are wrestling with the way to charge for the data aspects of GPRS. One *Regional Bell Operating Company* (RBOC) venture is planning an all-you-can-eat type service. For a flat rate, you can use as much data as you can with no added charges. Another is considering the use of the Japanese NTT DoCoMo model, whereby they charge a rate of the U.S. equivalent of $.0025 per packet of data sent. This constitutes a pay-as-you-go model. Still others are considering a pay-per-minute model that is similar to the telephony networks. Finally, some feel that the QoS or aggregate throughput is what should be charged at the end of the month. From experience in the field, the two that carry the most weight today are the DoCoMo model of charging per packet and the all-you-can-eat model.

GPRS is a different kind of service from those typically available on today's mobile networks. GPRS is a packet-switching overlay on a circuit-switching network. The GPRS specifications stipulate the minimum charging information that must be collected in the Stage 1 service description. These include destination and source addresses, usage of radio interface, usage of external packet data networks, usage of the PDP addresses, usage of general GPRS resources, and location of the mobile station. Because GPRS networks break the information to be communicated down into packets, at a minimum, a GPRS network needs to be able to count packets to charging customers for the volume of packets they send and receive. Today's billing systems have difficulty handling charging for today's nonvoice services. It is unlikely that circuit-switched billing systems will be able to process a large number of new variables created by GPRS.

GPRS call records are generated in the GPRS service nodes. The GGSN and SGSN may not be able to store charging information, but this charging information needs to be processed. The incumbent billing systems are often not able to handle real-time Call Detail Record flows. As such, an intermediary charging platform is a good idea to perform billing mediation by collecting the charging information from the GPRS nodes and preparing it for submission to the billing system. Packet counts are passed to a Charging Gateway Function that generates Call Detail Records that are sent to the billing system. The crucial challenge of billing for GPRS and earning a *return on investment* (ROI) on GPRS is simplified by the fact that the major GPRS infrastructure vendors all support charging functions as part of their GPRS solutions. Additionally, a wide range of other existing non-GSM packet data networks such as X.25 and *Cellular Digital Packet Data* (CDPD) are in place along with associated billing systems.

It may well be that the cost of measuring packets is greater than their value. The implication is that a per-packet charge will not occur because too many packets are present to count and charge. For example, a single e-mail application can generate tens of thousands of packets regularly. Therefore, the Charging Gateway Function becomes more of a policing function than a charging function. This lends credence to the idea that the network operators may tariff certain amounts of GPRS traffic at a flat rate and then only monitor whether these allocations are exceeded. If excess packets are used, then a value-added charge may occur.

This does not imply that the operators will offer the free Internet Service Provider model seen on the fixed Internet. Users do not pay a fixed monthly charge so the network operators rely on advertising sales on mobile portal sites to make money. A premium exists for the sake of being mobile and the costs associated with acquiring bandwidth dictates some form of charge-back system. Given the additional customer care and billing complexity associated with mobile Internet and nonvoice services, network operators would be ill-advised to reduce their prices in such a way as to devalue the perceived value of mobility.

The implementations by many vendors occurs on a Sun Enterprise Server (like the E250) hardware running some form of charging and accounting software. The Charging Gateway Function, shown in Figure 4-15, is composed of three main functional areas:

- **Billing Record Collector**   This entity collects billing records from the GPRS nodes. In the first GPRS release, only the GGSN Billing Record Collector was available.

- **Flow Aggregation Processor**   This entity aggregates several billing records produced during a PDP session (such as several Start and Stop

Figure 4-15
The Charging
Gateway Function.

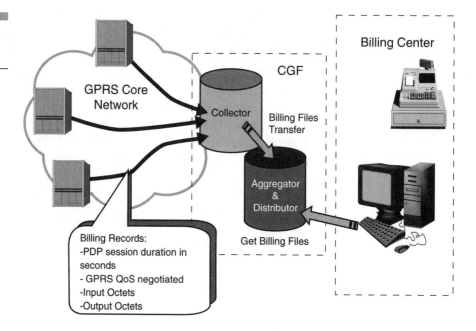

records due to time or volume conditions) into a single GPRS accounting record. The format for this aggregated record is called a *Network Accounting Record* (NAR). The billing records coming from different collectors cannot be aggregated (two different collectors are seen as two different billing streams by an aggregator).

■ **Flow Data Distributor** This entity is responsible for providing access to the GPRS accounting records to the customer billing system. The available interfaces are FTP, ASCII, or ASN.1 formats.

The architecture is typically distributed across two Sun servers for redundancy and robustness, but it is possible to have them running on a single server.

The Operations and Maintenance Center (OMC) and the Network Management Center (NMC)

In all large telecommunications networks, one of the critical components is the ability to maintain and manage the network. In GSM networks, similar functions are required that are local to the switches and base stations. The

Operations and Maintenance Center (OMC) is at this local level. In the over-all national or regional network, all the OMCs report to the *Network Management Center* (NMC).

OMC functions include

■ Events and alarms on all switching components and procedures

■ Fault management

■ Performance management

■ Security management

NMC functions include

■ Trunk route management

■ High-level alarms

■ OMC assistance

OMC Communication GPRS Domain Managers

GPRS has two major impacts on the Operations and Administration Systems, as shown in Figure 4-16:

■ On the OMC-R, the presence of the PCUSN causes an impact to the OMC.

■ A new domain manager is needed to manage the GPRS core network elements. We need a new OMC domain called *OMC-Data* (OMC-D).

PCUSN OAM Server

PCUSN management is performed at OMC workstation. The PCUSN configuration is performed at the workstation (OMC) and uses the PCUSN OAM server utilities. The A_{gprs} interface configuration is performed from the OMC workstations, but uses OMC-R server utilities. Finally, PCUSN alarms are sent towards the OMC-R through the PCUSN OAM server for alarm translation.

The OMC workstation is used for performance, fault management, and A_{gprs} configuration. This can all be done directly through an OMC-R Window. PCUSN configuration can be done from a UNIX background menu on the workstation screen. The interface is shown in Figure 4-17.

Figure 4-16
Domain
Management.

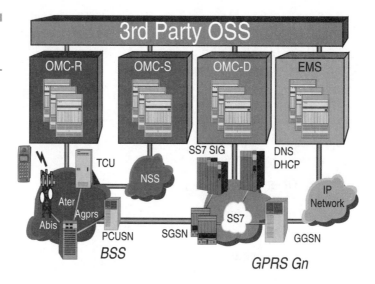

Figure 4-17
OAM Management.

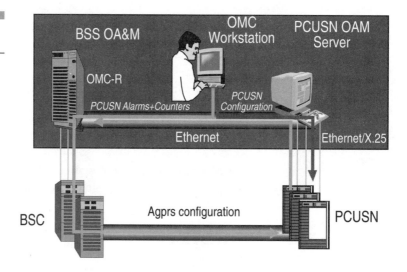

OMC-D Architecture

OMC-D device support provides for fault management, configuration management, performance management, and security management for the following devices as shown in Figure 4-18:

■ GGSN

■ SGSN

Figure 4-18
The GGSN.

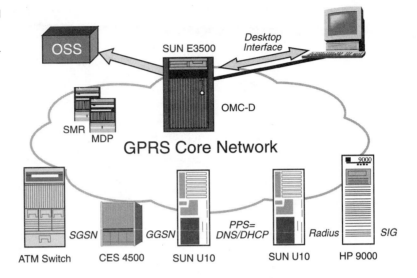

- ■ SIG
- ■ *Policy Servers* (PSs) (such as DNS/DHCP)
- ■ RADIUS

The OMC-D Core Management Client hosts many of the software releases for the management systems.

OMC-D Core Management Servers typically use Sun servers and are usually deployed in mated pairs (a primary-server operation is normal; a secondary-server operation is a warm backup).

A *Management Data Provider* (MDP) server collects raw performance data from the SGSN and other network devices to provide statistical performance records (by the external OSSs).

A *Service Management Reporter* (SMR) hosts the network performance information databases and provides an SQL interface to reporting tolls residing on the SMR client (PC).

Having looked at the various nodes and interfaces added to a GSM network in support of the GPRS services, the operators can actually get into the packet data business without major overhauls to their networks. The main investments are in software in many of the existing network components. The few actual hardware pieces are not so expensive as to prevent widespread acceptance and implementation of the data networks. In the next chapter the focus will leave the hardware components and emphasize the main procedures used within a network to efficiently use resources and to allocate radio services to the data user.

Main GPRS Procedures

Objectives

When you complete the reading in this chapter, you will be able to

- Describe the main components of the main functions of mobility management.
- Describe the mobile-initiated attach and detach procedures.
- Understand when the network performs an attach or detach procedure.
- Discuss the role of the SGSN in the procedures.
- Understand the three states of the mobile in GPRS.

Mobility Management (MM)

Before a *mobile station* (MS) can send data to a corresponding host, it must attach to a *Serving GPRS Support Node* (SGSN). An attachment procedure (GPRS attach) between the mobile station and the network is conducted and a *Temporary Logical Link Identifier* (TLLI) is assigned to the mobile station. Actually, the mobile uses the TLLI after the network assigns a *Packet Temporary Mobile Subscriber Identity* (P_TMSI). The mobile chooses the TLLI after being assigned the P_TMSI.

After attaching, one or more routing contexts for one or more *Packet Data Protocols* (PDPs) can be negotiated with the SGSN. Three *mobility management* (MM) states are related to a GPRS subscriber and each state describes the level of functionality and information allocated.

In idle state, the mobile station is not yet attached to the GPRS mobility management and a GPRS attach procedure must be performed. The conditions of the idle state are shown in Figure 5-1 when looking at the *radio resource* (RR) management.

In ready state, the mobile station is attached to GPRS mobility management (GMM) and is known in the accuracy of the cell. Each cell in a GSM network has its own *Cell Global Identity* (CGI), which enables the network to identify the mobile station by the cell. Each cell is associated with a *location area* (LA) in GSM, but in GPRS, the association is with the *routing area* (RA). The mobile station may receive and send data for all relevant service types. If the ready timer (for the mobile station or SGSN) expires, the mobile station will move to the standby state. Figure 5-2 shows the conditions for moving to and from the ready mode.

In the standby state, the subscriber is attached to the GPRS mobility management and is known in the accuracy of the routing area. The mobile

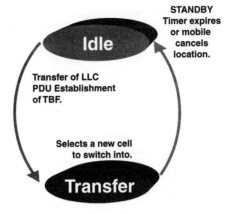

Figure 5-1
The parameters for
the idle state of
the mobile.

Radio Resource
State Machine

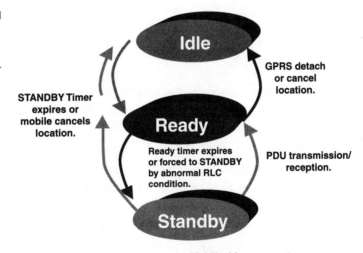

Figure 5-2
The MS in the ready
mode.

Mobility Management
State Model
of mobile station and SGSN

station performs a GPRS RA update and GPRS cell selection and reselection locally.

At this point, if the subscriber wants to request e-mail or a Web page, a PDP context must be activated in advance. If the standby timer (for the mobile station or SGSN) expires in this state, the mobility management contexts in both the mobile station and SGSN independently return to the

Figure 5-3
The PDP context
mode for the MS in
active or idle modes.

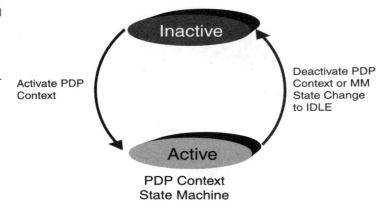

idle state and may be deleted. Figure 5-3 shows the conditions of the PDP context management mode.

GPRS Attach

A GPRS mobile station is not reachable or known by the network until the mobile station performs the attach procedure and switches into the ready mode. To attach to the network (actually the SGSN), the mobile station provides its identity and indicates which type of attach procedure is to be performed:

- **GPRS attach** Needs the mobile station's P_TMSI and the *routing area identity* (RAI) where the mobile is located
- **IMSI attach** Specific to GSM, but may be performed via GPRS if a TMSI or P_TMSI are not already assigned to the mobile station
- **IMSI/GPRS attach (for class A and B mobile stations)** Will be possible in a later release of GPRS

In the ready state,

- Both the mobile station and the SGSN have established mobility management contexts for the subscriber's *International Mobile Subscriber Identity* (IMSI), which is the primary key to the GPRS subscription data stored in the *Home Location Register* (HLR).
- The mobile station may send and receive data *protocol data units* (PDUs) that are nothing more than packets.

- The mobile station may also activate or deactivate PDP contexts (data addresses) with the network. A mobile station may have many active PDP contexts simultaneously.

- The mobile station listens to the GPRS *Packet Common Control Channel* (PCCCH) and may also use *discontinuous reception* (DRX). Discontinuous reception means that the mobile station will only use resources when data is present to receive (in the form of packets). Other times, the always-on condition is not using any radio resource.

The mobility management remains in the ready mode until the ready timer expires and the mobile station then moves to the standby mode. Figure 5-4 shows the process of conducting the GPRS attach as the mobile station initiates the request.

GPRS Attach Scenario

Using the previous statements, we can then see the procedure that the mobile station uses to attach to the GPRS network. The mobile station wants to initiate a packet data session, (for example, access the Internet or check e-mail) from a wireless network (vis-a-vis GPRS). To do this, the

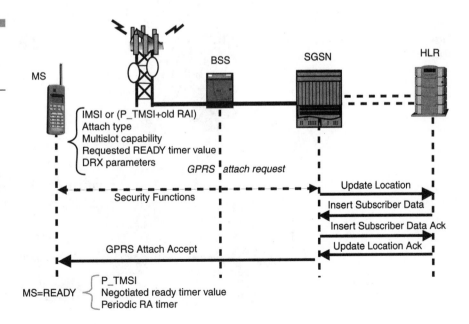

Figure 5-4
GPRS attach procedure to move to the ready mode.

mobile station must first attach itself to the wireless network—the SGSN to be more specific. Four steps are involved in the attach process, as shown in Figure 5-5:

1. The mobile station sends an attach request with its identity (P_TMSI or TMSI) to the SGSN. This message will also contain a *Network Service Area Point Identifier* (NSAPI), which is specific to a particular network application at the mobile station. The *Subnetwork-Dependent Convergence Protocol* (SNDCP) layer uses this NSAPI to communicate with the network application. Several NSAPIs may be associated with an individual mobile station. One may be for an Internet browser, whereas another could be an e-mail service.

2. The SGSN verifies whether the user is authorized and authenticated for that particular service by checking with the HLR entry for the mobile station.

3. After authorization, the SGSN sends back a reply to the mobile station with a TLLI. The TLLI is specific to the mobile and is used by the *Logical Link Control* (LLC) layer in the protocol stack. The purpose of this TTLI is to provide a temporary ID to the mobile station, which can be used for data communication.

4. A database is maintained at the SGSN that maps the mobile identity with the TLLI assigned to it. The NSAPI is associated with and the

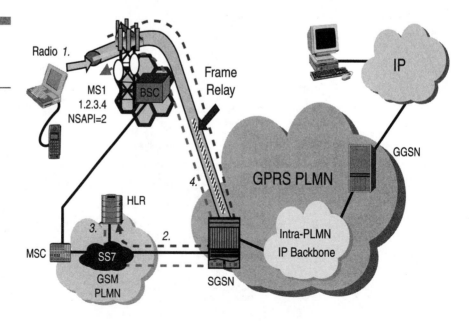

Figure 5-5
The steps in performing a GPRS attach.

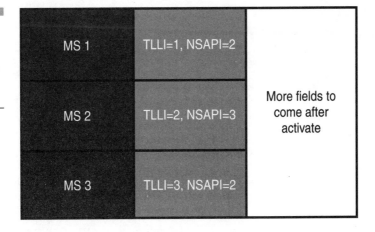

Figure 5-6
Table updates
correlating the
mobile station to the
SGSN for attach
procedures.

MS 1	TLLI=1, NSAPI=2	
MS 2	TLLI=2, NSAPI=3	More fields to come after activate
MS 3	TLLI=3, NSAPI=2	

quality of service (QoS) subscription parameters required by the application.

The table of entries for the SGSN correlations of attached mobiles shown in Figure 5-6 is updated after the PDP activation phase.

Mobile Station-Initiated GPRS Detach

To move from the ready state to the idle state, the mobile station initiates a GPRS detach procedure. The results of the GPRS detach function is that the SGSN may delete the MM and the PDP contexts; the PDP contexts are actually deleted in the *Gateway GPRS Support Node* (GGSN). The mobile station detaches by sending a detach request (detach type or a switch off) message to the SGSN. The detach type may include such things as detach for GPRS purposes only, IMSI detach (a GSM function), or both.

Detach Type—GPRS-Only, IMSI-Only, or Combined

If GPRS detaches, the active PDP contexts in the GGSN regarding this particular mobile station are deactivated by the SGSN sending a Delete PDP Context Request message to the GGSN. The GGSN acknowledges with a

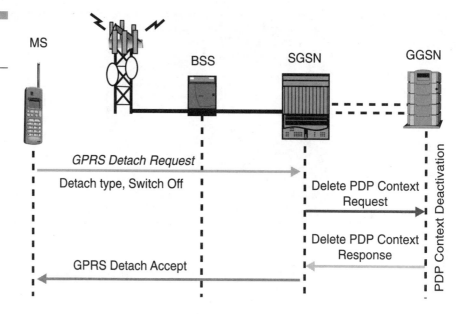

Figure 5-7
MS-initiated GPRS
detach procedures.

Delete PDP Context Response. If the switch off indicates that the detach request is not due to a switch-off situation, the SGSN sends a GPRS Detach Accept message to the mobile station. Figure 5-7 shows the mobile station-initiated detach procedures, where the steps indicate the type of detach indicated.

Network-Initiated GPRS Detach

When necessary for the network to detach a mobile station, the SGSN informs the mobile station that it has been detached by sending a Detach Request (Attach Indicator) to the mobile station. The Attach Indicator indicates if the mobile station is requested to make a new GPRS attach and perform PDP context activation procedures for the previously activated PDP contexts. If so, the GPRS attach procedure is initiated when the GPRS detach procedure is complete. Figure 5-8 shows the network-initiated detach procedure.

The active PDP contexts in the GGSN are deactivated by the SGSN. The mobile station sends a GPRS Detach Accept message back to the SGSN anytime after the Detach Request message.

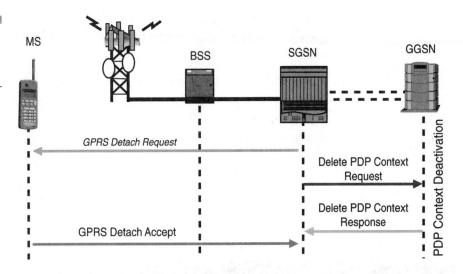

Figure 5-8
The network-initiated GPRS detach procedures.

Activating a PDP Context for Packet Routing and Transfer

Before data can be sent or received, a PDP context (a data address) must be activated (created for the mobile station). The PDP context is used for routing purposes inside the GPRS network. A GPRS subscription contains several PDP addresses and an individual PDP context is maintained in the mobile station, SGSN, and GGSN for every PDP address. All PDP contexts for a subscriber are associated with the same MM context for the IMSI of the subscriber. It is possible to inquire and/or set the following parameters:

■ Requested QoS such as the peak bit rate, mean bit rate, delay requirements, service precedence, and reliability level expected

■ Data compression or no data compression such as using V.42 bis data compression on the payload

■ Whether or not to use TCP/IP header compression

■ PDP address and type requested (particularly if an IP or X.25 address are static)

Each PDP context can be either active or inactive, as seen earlier in Figure 5-3, and three PDP context functions are available—activate, deactivate, or modify:

- The mobile station is responsible for activation and deactivation.
- GGSN is responsible for activation (for incoming packets) and deactivation.
- SGSN is responsible for modification.

The three functions are only meaningful at the *Network Subsystems* (NSS) level and do not directly involve the *Base Station Subsystems* (BSS). A mobile station in standby or ready state can initiate activation or deactivation at anytime to activate the PDP context in the mobile station, the SGSN, or the GGSN.

GPRS Context Activation—Scenario

Using a play on the PDP context activation, a mobile station has attached itself to a SGSN in the GPRS *Public Land Mobile Network* (PLMN). The mobile station has been assigned a TLLI that the wireless network knows. However, the external network nodes (IP or X.25) do not yet know of the mobile station. Therefore, the mobile station must initiate a PDP context with the GGSN.

Both the SGSN and the GGSN are identified by IP addresses. A many-to-many relationship exists between the SGSN and the GGSN. Multiple tunnels (used for secure data transfer between the SGSN and the GGSN) may exist between a pair of GGSNs, each with a specific *tunnel identifier* (TID). Four steps are involved in the activation process, as shown in Figure 5-9:

1. The mobile station sends a PDP context activation request to the SGSN.

2. The SGSN chooses the GGSN based on information provided by the mobile station and other configurations and requests the GGSN to create a context for the mobile station. The SGSN will select a GGSN that serves the particular type of context needed (such as one for IP network access and one for X.25 access)

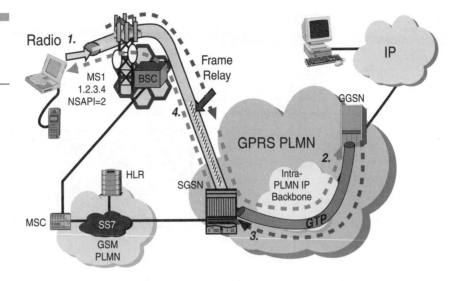

Figure 5-9
The steps in
activating a PDP
context.

3. The GGSN replies to the SGSN with the TID information. It also updates its tables wherein it maps the TID and the SGSN IP addresses with the particular mobile associated with them.

4. The SGSN sends a message to the mobile station informing it that a context has been activated for that particular mobile. The SGSN also updates its tables with the TID and the GGSN IP address with which it has established the tunnel for the mobile.

Figure 5-10 shows the table entries for the SGSN and the GGSN.

Mobile-Initiated PDP Data Protocol Context Activation

The PDP context activation aims to establish a PDP context between the mobile station and the network, as shown in Figure 5-11. It may be performed automatically or manually depending upon the manufacturer's implementation and configuration. The mobile station first sends an Activate PDP Context Request message that contains the following:

- NSAPI
- PDP type

Figure 5-10
The SGSN and
GGSN tables.

SGSN Table	MS 1	TID 1	GGSN IP
	MS 2	TID 2	GGSN IP
	MS 3	TID 3	GGSN IP

GGSN Table	MS 1	TID 1	SGSN IP
	MS 2	TID 2	SGSN IP
	MS 3	TID 3	SGSN IP

Figure 5-11
The mobile-initiated
PDP activation.

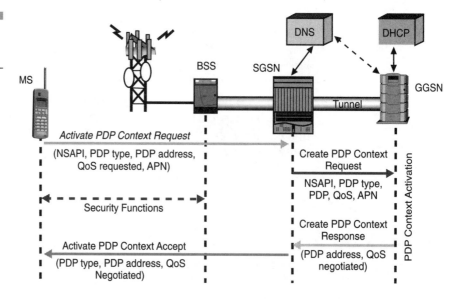

- PDP address, whether it is a static or dynamic address (IP address)

- Requested QoS (best effort is all that is currently available, but will get to specific QoS in the future)

- *Access Point Name* (APN) (optional) to select a certain GGSN, either the IP address or logical name is used

- PDP configuration options

The mobile station only exchanges messages with the SGSN, which acts as a relay to the GGSN. The SGSN performs the following:

- Check the subscription data that was stored in the SGSN during the GPRS attach to determine if the mobile station is able to activate the PDP address.
- Insert the NSAPI along with the GGSN address in its PDP context.
- Return an Activate PDP Context Accept message to the mobile station.
- Become ready to route PDP packets (PDUs) between the GGSN and the mobile station.

Continuing the process of the mobile-initiated PDP context activation, and using Figure 5-12 as a guide, the next steps are considered. When the SGSN receives an APN from the mobile station, it checks to see if the APN ends with .gprs. If it does, the last three labels of the APN that represent the APN Operator Identifier are removed and the remaining labels that make up the APN Network Identifier are used for comparison with the subscription record.

In the second step, when the SGSN has validated the mobile station parameters, it must determine which GPRS gateway to select for the given APN. Therefore, it sends a request DNS query in which it provides the APN Network Identifier and APN Operator Identifier. The DNS responds with a list of the available GGSNs to use, in a preferred order. After that, the

Figure 5-12
The mobile-initiated
PDP context
activation continued.

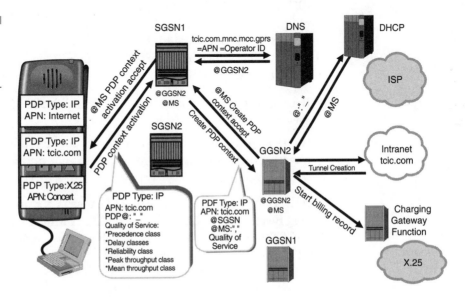

SGSN knows which GGSN to select and sends the activated PDP context to the correct serving gateway and opens a GTP tunnel if it has not already done so. A tunnel between the SGSN and the GGSN is identified by the TID.

Upon receipt of the Create PDP Context Request message, with the help of the APN, the GGSN needs to determine the following:

- The access mode used on the external network (in this example, we consider an intranet access with a secure tunnel)

- The IP address allocation type (in this example, we use the *Dynamic Host Configuration Protocol* [DHCP] server)

Finally, the GGSN starts the billing records for this PDP context. It also returns an accept message to the SGSN, including the IP address for the mobile station.

Network-Initiated Packet Data Protocol Context Activation

The network may also initiate a PDP context if data arrives for a mobile user who has not established an address already. Figure 5-13 shows this network-initiated context activation. When receiving a PDP PDU, the GGSN determines if the network-initiated PDP context activation procedure has been initiated. The GGSN may send a Send Routing Information for GPRS (IMSI) message to the HLR (via the SGSN). The HLR returns a Send Routing Information for GPRS ACK (the information contained includes the IMSI, SGSN address, and cause) message to the GGSN:

- If a request can be served, the HLR includes the IP address of the serving SGSN.

- If a request cannot be served, the HLR only includes cause to indicate the reason for the negative response (cannot find network address and so on).

If the SGSN address is present and cause is not present or equal to a No Paging Response, the GGSN sends a PDU Notification Request message to the SGSN indicated by the HLR, which acknowledges it by sending a PDU Notification Response message.

Figure 5-13
The network-initiated
PDP context
activation.

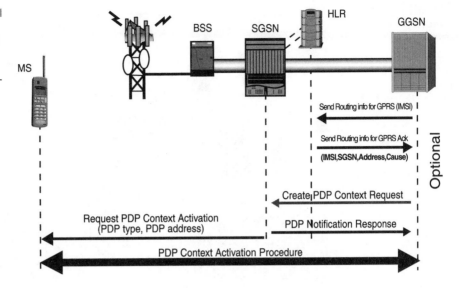

The SGSN sends a Request PDP Context Activation (PDP type, PDP address) message to request that the mobile station activate the indicated PDP context, using the PDP context activation procedure (same as the mobile station-initiated).

GPRS Data Transfer from the Mobile Station

After attaching to the SGSN and activating a PDP context, the mobile station is now known to the external *packet data network* (PDN) and can send and receive information to and from the networks. Now a user application at the mobile station is going to generate IP or X.25 packets. The packets contain a source address, a destination address, and information. The flow of the packets is listed in the following steps and shown in Figure 5-14:

1. A logical link exists between the SGSN and the mobile station. The link is identified by the TLLI specific to the mobile station. A table exists in the mobile station that holds the mapping information of the mobile to the TLLI and the associated NSAPI. The SNDCP layer takes

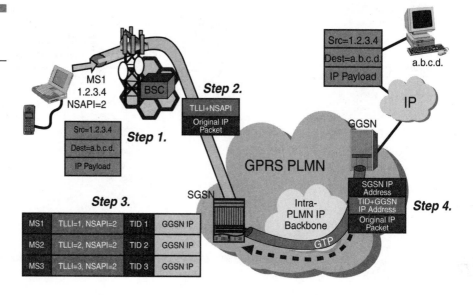

Figure 5-14
GPRS Data transfer
from mobile.

the original IP packet and adds header information containing the
TLLI and the NSAPI information. These packets are then sent to
the SGSN.

2. The table at the SGSN also holds mapping information of the TLLI and
   NSAPI to the corresponding TID and GGSN IP addresses. At the
   SGSN, the header that contains the TLLI and NSAPI is removed and a
   GTP header containing the TID and the GGSN IP address is put in its
   place.

3. The packets are sent to the GGSN in the IP format with the IP address
   of the SGSN as the source address and the GGSN IP address as the
   destination. The TID is also part of the IP datagram (packet).

4. At the destination (GGSN), the header is stripped off and the original
   IP or X.25 packet is obtained. This packet can now be routed to its
   destination from the destination address field of the packet.

GPRS Data Transfer to the Mobile

Data transfer to the mobile station is similar to the data transfer from the
mobile station. Figure 5-15 shows the steps, which follow this sequence:

Figure 5-15
GPRS data flow to
the MS.

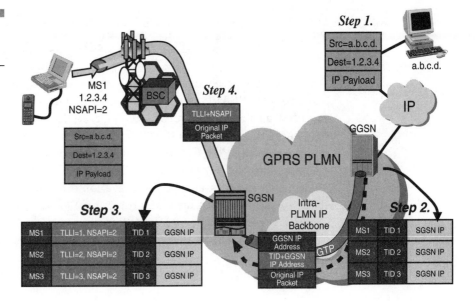

1. Packets from the external network reach the GGSN. The GGSN looks up the tables to determine the particular SGSN address and TID for the mobile station that is the intended recipient of the packet.

2. The GGSN forms an IP datagram (packet) with the GGSN IP address as the source address and the SGSN IP address as the destination address and the original IP packet inside. The packet also contains the TID.

3. The SGSN maps the TID and SGSN to the corresponding TLLI and NSAPI values at the table entries. At this point, the SGSN knows where the mobile station is and to which network application it must route the packets.

4. The SGSN takes the original IP packet, adds a header with the NSAPI and TLLI, and forwards it to the mobile station. The SNDCP layer at the mobile station strips off the header and sends the packet to its associated network layer application.

This sequence assumes that an active PDP context is open. In the event that no PDP context is established for the mobile station, the need may arise for the network to initiate a PDP context. First, a context is established between the GGSN and the SGSN to which the mobile station is attached. Then, the flow of events is as previously listed.

Mobile-Initiated Packet Data Protocol Context Deactivation

To initiate this procedure, the mobile station sends a Deactivate PDP Context Request (NSAPI) message to the SGSN and security functions may be executed, as shown in Figure 5-16. The SGSN sends a Delete PDP Context Request (TID) message to the GGSN, which removes the PDP context and returns a Delete PDP Context Response (TID) message to the SGSN.

If a mobile station were using a dynamic PDP address, the GGSN would release this address and make it available for subsequent activation by other mobile stations. The SGSN returns a Deactivate PDP Context Accept (NSAPI) message to the mobile station.

At GPRS detach, all PDP contexts for the mobile station are implicitly deactivated.

Network-Initiated Packet Data Protocol Context Deactivation

When the GGSN initiates the PDP context deactivate procedure as shown in Figure 5-17, it sends a Delete PDP Context Request (TID) message to the

Figure 5-16
Mobile-initiated PDP context deactivation.

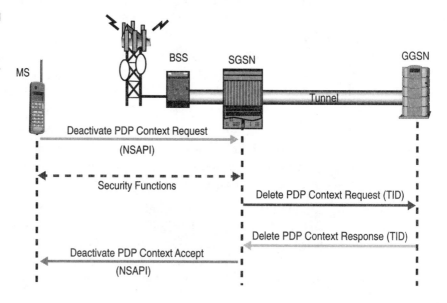

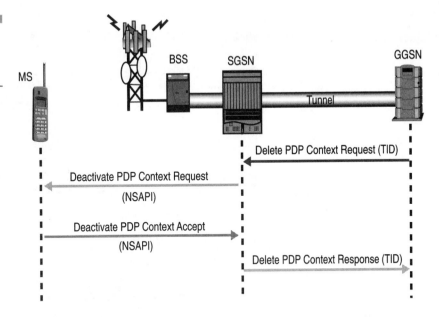

Figure 5-17
Network-initiated
PDP context
deactivation.

SGSN, which sends a Deactivate PDP Context Request message (NSAPI) to the mobile station. The mobile station removes the PDP context and returns a Deactivate PDP Context Accept (NSAPI) to the SGSN.

The SGSN returns a Delete PDP Context Response (TID) message to the GGSN. The SGSN may not wait for a response from the mobile station before sending the Delete PDP Context Response message.

Security Functions

Security in GSM and GPRS networks is based on the following two primary techniques, which are shown in Figure 5-18:

- Authentication
- Ciphering (encryption)
- authenticating the user

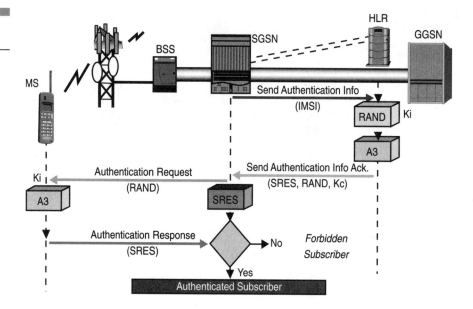

Figure 5-18
Security functions.

Authentication

The *Authentication Center* (AuC) is responsible for generating a set of parameters known as triplets. A triplet consists of a

- *Cipher Key* (K_c)
- *Random Number* (RAND)
- *Signed Response* (SRES)

The RAND is a randomly generated number from a number pool containing 2^{128} numbers. The RAND, coupled with the *Identification Key* (K_i), is used to calculate K_c and SRES. K_i is a secret number allocated on a persubscriber basis and is only held at the AuC and is based on the *Subscriber Identity Module* (SIM) card. Measures are taken to ensure that the K_i cannot be read from the SIM card. K_i is never transmitted over the network. Authentication procedure, based on the GSM, performs the selection of a ciphering algorithm. The SGSN may store the authentication triplets of the mobile station after detaching from the GPRS. If it does not have the previously stored authentication triplets, they can be obtained from the HLR.

Ciphering

The mobile station starts ciphering after sending the Authentication Response to the SGSN. On receipt of a valid response message, the SGSN starts ciphering. Ciphering is used over the air interface following the authentication procedure to provide security for voice and data traffic. *Algorithm 5* (A5) is used with K_c and current *Time Division Multiple Access* (TDMA) frame number as inputs to generate a ciphering code. The mobile station calculates K_c from the RAND and K_i and stores it on the SIM. The BSS is given K_c by the *Visitor Location Register* (VLR) or SGSN. In the uplink direction, the mobile ciphers the data and the BSS deciphers it. A similar process takes place on the downlink.

The cipher key is different in the uplink and downlink direction. The TDMA frame number changes approximately every 4.6 ms (a TDMA frame period) and is not repeated for 3.5 hours, making it difficult for the cipher code to be cracked. Some countries allow ciphering as an option, others forbid it.

The network also has the option to start ciphering without authentication.

Web Access

To achieve a Web access, the mobile station first performs a GPRS attach procedure to become ready and then the PDP context activation to establish communications with an Internet host, as shown in Figure 5-19.

The SGSN encapsulates the outgoing data and routes the packets to the appropriate GGSN, where they are sent to the Internet. Inside this network, PDN-specific routing procedures are applied to send the packets to the corresponding host. In the other direction, the incoming data are carried out to the *Packet Control Unit Support Node* (PCUSN), which first establishes a downlink radio channel with the mobile station.

Using the TTLI for the mobile station, the PCUSN notifies the mobile station of the channels (and the *uplink state flags* [USFs]) over which the data will be transferred using the Downlink Packet Resource Assignment message. The procedure continues with the data transfer to the mobile station.

The main functions of the GPRS procedures create the means for the mobile station to attach or detach from the network. Many other procedures

Figure 5-19
Web access on GPRS.

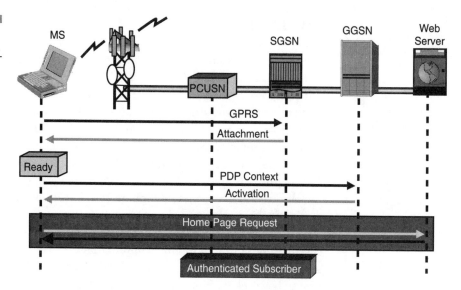

such as the authentication and ciphering functions support these procedures. Moreover, DNS and DHCP services are provided on an as-needed basis. From this point, the next step is to allocate the radio resources and establish a data flow. Chapter 6, "Radio and MS-PCUSN Interfaces," delineates these resource assignments and the process of preparing to send and receive the data.

Radio and MS-PCUSN Interfaces

Objectives

Upon completion of this chapter, you should be able to

- State the different air interface requirements used.
- Understand the resource allocation methods.
- Describe how the timing advance is used in GPRS.
- Explain the coding schemes used in GPRS.
- Describe the function and role of the PCUSNs.

Radio Link Control/Medium Access Control and Radio Frequency Layers

This chapter focuses on the radio interface (the interface between the *mobile station* (MS) and the *Base Transceiver System* (BTS), which is functionally the *Global System for Mobile Radio Frequency* [GSM RF] layer) and on the mobile station to the *packet control unit* (PCU) interface, which is the *Radio Link Control / Medium Access Control* layers (RLC/MAC layer).

To start the whole process off, the radio interface corresponds to the software in Layer 1 (GSM RF layer) between the mobile station and the BTS using the OSI model as the base reference, which is shown in Figure 6-1. The GSM RF layer manages the physical link between the mobile station and the *Base Station Subsystem* (BSS) (the combined BTS and BSC). This layer corresponds physically to the *Channel Codec Unit* (CCU) inside the BTS. Sometimes the reference materials show this layer divided into two sublayers, including the

- **Physical RF layer** The physical RF layer corresponds to the modulation and demodulation tasks, similar to GSM current techniques (GMSK modulation and Viterbi demodulation), but plans call for this to change in further evolutions of GPRS with the introduction of a more spectrum-efficient modulation (*Enhanced Data rates for GSM Evolution* [EDGE]).

- **Physical link layer** The physical link layer provides information transfer over a physical channel on the radio interface. It provides channel-coding functions (*forward error correction* [FEC]), interleaving,

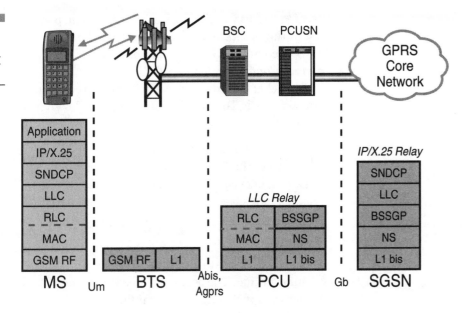

Figure 6-1
The protocol stacks
for the RLC and MAC
layers.

radio channel measurement functions (received quality and signal levels, timing advance measurement, physical link congestion detection), and radio management procedures (cell selection/reselection, power control, and *discontinuous reception* [DRX]). It does not perform ciphering (which is handled by the *Logical Link Control* [LLC] layer).

The MS-to-PCU interface corresponds to the supervision level of the radio communications vis-a-vis RLC/MAC layer. This layer manages the logical link between the BSS and the MS.

- RLC
- MAC

Packet Logical Channels

GPRS uses some GSM broadcast channels for frequency tuning (*Frequency Control Channel* [FCCH]) and synchronization (*Synchronization Channel* [SCH]). However, for other purposes, specific new packet logical channels are defined, which are carried by a packet-switched channel *Packet Data Channel* (PDCH). PDCH is the generic name for the physical channel allocated to carry packet logical channels. The packet (logical) channels are used in the *uplink* (UL) or *downlink* (DL) direction. Actually, unlike GSM, GPRS does not have a real duplex channel, except for the *Packet Timing Advance Control Channel* (PTCCH).

The packet channels are classified in three families (the direction of flow is shown as UL, DL, or UL/DL). These are shown as the logical channels in Figure 6-2:

■ The *Packet Common Control Channels* (PCCCH) are very similar in makeup to the *Common Control Channel* (CCCH). When not allocated in a cell, packet transfer can be initiated by the CCCH. The PCCCH is made up of several logical channel functions consisting of

■ *Packet Random Access Channel* (PRACH UL) is used for random access. The PRACH is an uplink-only function used by the mobile station to initiate an uplink transfer for sending data or signaling information. The access burst used on the PRACH is also used to obtain any timing advance information. The mobile station to transmit the initial packet channel request uses the PRACH. It is the only request for short access or one-phase access. In the case of two-

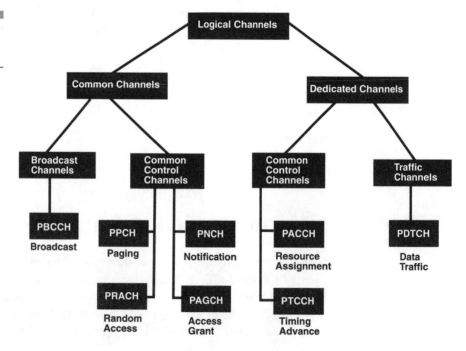

Figure 6-2
The logical channels assigned in GPRS.

phase access, a second request follows (Packet Resource Request on *Packet Associated Control Channel* [PACCH]).

Two types of random access bursts may be transmitted on the PRACH: an 8-information-bit random access burst or an 11-information-bit random access burst called *extended packet random access burst* (only the 8-information bits access burst format can be used on RACH in the GSM system or in the GPRS system if the PRACH is not used). The 11-bit format contains additional p bits to manage the request priorities that the 8-bit format does not contain.

The access burst type that the MS uses in one cell is indicated through the parameter ACCESS_BURST_TYPE broadcast in the *Packet System Information* (PSI) on the PCCCH (if it exists). If the PRACH also does not exist, then the 8-bit format is the only one that the MS can use on the RACH.

- *Packet Paging Channel* (PPCH DL) is used for paging the mobile station prior to a mobile station receiving (terminating) data transfer. The PPCH uses paging groups to provide DRX and follows the same predefined rules as the Paging Channel (PCH) in GSM. Mobile stations in both circuit- and packet-switching modes can be paged, although this is only applicable for GPRS mobile stations class A and B. Beyond that, a mobile station engaged in a packet-switched transfer can be paged on a PACCH.

  PPCH is used to page the mobile station in standby mode in its routing area for a mobile telephone call before the assignment of the *Packet Data Traffic Channel* (PDTCH DL) on the *Packet Access Grant Channel* (PAGCH) of the cell where the mobile station is located. PPCH is transmitted on a normal burst.

- PAGCH DL for Immediate Assignment is a downlink-only channel used during the setup of a packet transfer to send Resource Assignment messages. If the mobile station is currently involved in packet transfer, then the Resource Assignment messages can be sent on the PACCH.

  In one-phase access or short access, PAGCH assigns several blocks. In two-phase access, PAGCH assigns one single block, on which the mobile station will send its Packet Resource Request (on the PACCH) to get several blocks on a Packet Assignment Request (using the PACCH as well). PAGCH is transmitted on a normal burst.

- *Packet Notification Channel* (PNCH DL) is used for *Point-to-Multipoint-Multicast* (PTM-M) notification. In GPRS Phase II, it is a downlink-only channel that sends PTM-M notification to a group of mobile stations prior to the PTM-M packet transfer actually taking place. This notification is in the form of a Resource Assignment message.

■ The *Packet Broadcast Control Channel* (PBCCH) is DL only. It broadcasts packet data system information and follows the same predefined rules for mapping onto the physical channels as the *Broadcast Control Channel* (BCCH) in GSM. The existence of the PBCCH is indicated on the BCCH and if it is not allocated, the packet system is contained in the BCCH control messaging system.

■ The *Packet Traffic Channels* (PTCHs), comprising

- PDTCH UL or DL is used for data traffic. This channel is used for data transfer and mapped directly onto one of the physical channel (*Time Division Multiple Access* [TDMA] time slots). These are temporarily dedicated to one mobile station or a group of mobile stations. One mobile station may use multiple PDTCHs in parallel for individual packet transfer. Up to eight PDTCHs, each with different time slots, may be allocated to one mobile station at one time or to a group of mobile stations in the case of PTM-M.

- PACCH UL or DL is used for control signaling. This channel is a dedicated control channel function that conveys signaling information related to a mobile station. This includes acknowledgements and power control information. It also carries Resource Assignment and Reassignment messages consisting of the assignment of a capacity for a Packet Data Traffic Channel and for further occurrences of the PACCH. The PACCH shares resources with the PDTCH currently assigned to one mobile station. A mobile station currently involved in packet transfer can be paged for circuit-switched services on this PACCH.

- PTCCH UL and DL is a dedicated control channel for *timing advance* (TA) updates. The uplink portion of the Timing Advance Common Control Channel uses random access bursts to provide an estimation of timing advance. The downlink portion of the timing advance transmits timing advance information to several mobile stations. Typically, one PTCCH downlink is paired with several PTCCH uplinks. In timing advance for GSM, the receiving nodes estimate the appropriate time for the reception of the bursts from the mobile station. The cell sizes tend to be limited to a maximum radius of

35 km. The maximum timing advance is 63 bits (0 to 63). The duration of a single bit is 3.69 ms. Because the path to be equalized is a two-way service, the maximum physical distance between the BTS and a mobile station is half the maximum delay, or 70 km / 2 = 35 km. The random access burst can accommodate a maximum delay over a distance of 75.5 km; that way the bursts can appear at the BTS receiver with a high possibility that another mobile's normal burst will not cover them. With larger cells, this cannot be assured. Clearly, the overall system performance is partially based on distances from the BTS. The variability that creeps into the network includes the mobility of the users, so that they are at different locations and distances from the BTS. As a result, the propagation delay is different. Moreover, the moving target in a mobile station creates a very changing reception from the mobile. Each of these conditions must be met in order to deliver reasonable quality (speech or data) in a GSM network.

Some problems are created as stated at different distances from the BTS for each mobile communicating with the BTS. Delay times in round-trip propagation and the attenuation of the signal are different.

To solve this problem, timing advance (up to 63 bit times) can be used to compensate for various distances and delays in the air. In effect, this creates an overall average of delay that the systems can deal with. By having a station transmit a few bit times early, the BTS is able to compensate for the arrival (either early or late) and address the proper time slot.

To illustrate how to compensate for the variable delays and the timing that is required, the timing advance is used, as shown in Figure 6-3. In this scenario, a normal burst for two different mobiles is being sent through the airwaves. The first mobile's burst arrives in time and is slotted into time slot number 6 properly. However, transmitter number 2's burst arrives late, and overlaps time slots. It is partially in the time slot for number 7 and a portion of the burst falls into the time domain for time slot number 0.

The base station notices that the burst is arriving late. Therefore, it sends a directive on the downlink control channels to the mobile. The directive tells the mobile to transmit its data earlier so that the delay is compensated for and the burst arrives within the appropriate timing, as shown in Figure 6-4. Now the transmitted

Figure 6-3
The data bursts are
arriving late and
overlapping a
time slot.

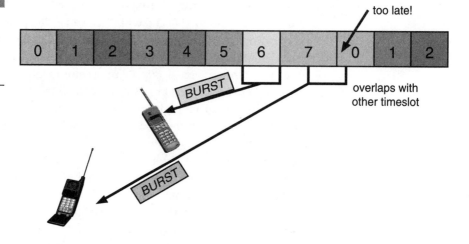

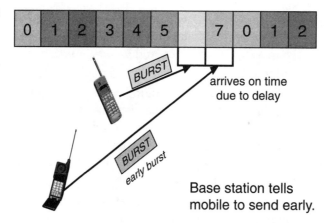

Figure 6-4
TA solves the
problem.

normal burst will arrive in time slot domain for slot number 7.
Timing advance can increase the transmission by as much as 63 bit
times (63×3.69 microseconds).

In GPRS Phase I, only the PTCHs will be used. The CCCHs of GSM will
be used instead of the PCCCHs and the BCCH of the GSM will be used
instead of the PBCC. All the necessary parameters for GPRS access will
be broadcast on the BCCH using a *System Information type 13* (SI 13)
message.

Packet Logical Channels—PDCH Allocation

The time slot configuration is declared for each time slot at the *operations maintenance center for radio* (OMC-R), initially found in GPRS Phase I. As shown in Figure 6-5, some time slots are reserved for the GSM system only (circuit-switched time slot); others are reserved for the GPRS system only (packet-switched time slot: PDCH); and some are used for a mix of each (first come first served). Future GPRS phases will likely have a dynamic time slot configuration, so that all the channels may be allocated either for a circuit-switched logical channel or for a packet-switched logical channel, based on the capacity-on-demand principle (TCH/PDTCH configuration of the time slot at the OMC). The PDCH may be temporarily allocated (they share the same physical resource as circuit-switched services) due to their fast release.

A cell supporting GPRS may allocate resources on one or several physical channels in order to support the GPRS traffic. The physical channels shared by GPRS mobile stations are taken from a common pool of physical channels available within the cell. This allocation of physical channels to switched services and GPRS is carried out somewhat dynamically according to the capacity-on-demand principle.

Figure 6-5

The sharing of time slots between GSM and GPRS.

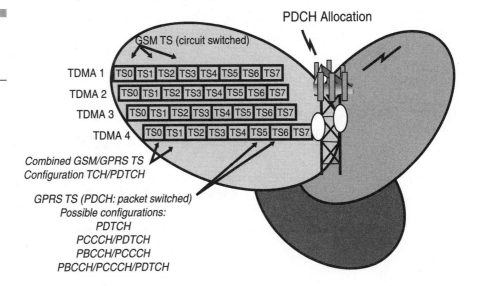

The capacity-on-demand principle means that PDCHs do not need to be permanently allocated to support GPRS and the network allocates available channels as required. Common control signaling required by GPRS in the initial phase of packet transfer is conveyed on the PCCH, when allocated to the network, or on the CCCH using GSM specifications. This saves on specific GPRS capacity for the *Public Land Mobile Network* (PLMN) operator. If the last available PCCH is allocated, then the mobile stations will perform a cell reselection.

At least one PDCH, acting as the master channel, carries the PCCH as well as the PDTCH and PACCH. Other PDCHs, acting as slave channels, are used for data transfer and dedicated signaling. The possible configurations for the packet-switched channels (PDCHs) are

- PDTCH
- PCCCH + PDTCH
- PBCCH + PCCCH
- PBCCH + PCCCH + PDTCH

These configurations will result in different packet channel multiplexing on the same PDCH. The fast release of the PDCH is an important feature that enables the dynamic sharing of the physical *radio resources* (RRs) between packet- and circuit-switched services. To enable this, three PDCH release options are available:

- Wait for assignments to terminate on that PDCH.
- Individually notify all users who have assignments on that PDCH.
- Broadcast the notification about deallocation.

Packet Logical Channels— Multiframe Structure

The packet channels carry either RLC data blocks or RLC/MAC control blocks (except PRACH and PTCCH UL, which use an access burst instead of normal bursts). Each of these radio blocks is mapped after channel coding and interleaving onto four radio frames (called radio blocks because they carry logical radio blocks).

The mapping in time of the packet logical channels carried by the same PDCH is defined by a multiframe structure. The multiframe structure for PDCH consists of a cycle of 52 successive TDMA frames, divided into 12 blocks (of four frames each), and 4 idle frames. The multiplexing of the packet channels on a PDCH is not fixed like in the GSM system. It is managed by some parameters and the following block order: B0, B6, B3, B9, B1, B7, B4, B10, B2, B8, B5, and B11. For example, if the cell has four PBCCH blocks, those will be carried by the blocks B0, B6, B3, and B9 (on the same time slots indicated in SI 13 on BCCH).

The idle frames are used by the mobile station for signal measurements and *Base Station Identity Code* (BSIC) decoding on the SCH of neighboring cells (idle 2 and 4) or for TA update (sending an access burst on PTCCH uplink in idle 1 or 3 and receiving an RLC/MAC control block on PTCCH DL in idle 1 and 3 of two successive multiframes, which equals four frames in total).

The multiframe for PDCH consists of 52 TDMA frames, divided into 12 blocks of 4 frames (radio blocks) and 4 idle frames. This multiframe, shown in Figure 6-6, can be seen as two 26-frame multiframes on the GSM network, numbered from 0 to 51. The multiframe has a duration of 240 ms and 25.5 multiframes are counted as a superframe.

Figure 6-6

The 52 multiframes in GPRS.

Cycle of 52 TDMA frames divided in:
- Twelve radio blocks B0-B11 (of 4 consecutive frames)
- Four idle frames (X)

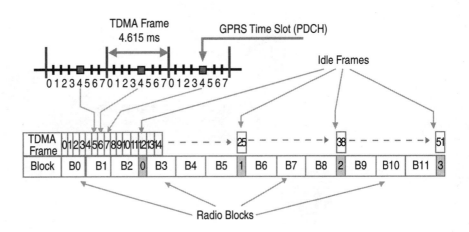

Packet Broadcast Control Channel (PBCCH)

The presence of a PBCCH channel in the cell is indicated by a PBCCH channel description in the SI 13 broadcast on the BCCH (providing GPRS-specific information). If the mobile station receives a SI 13 message without any PBCCH description, it will assume that PBCCH is not present in this cell (this was the case in GPRS Phase I), and the only PDCHs used for GPRS are the PDTCH and the associated control channels—PACCH and PTCCH. Figure 6-7 shows a representation of the PBCCH.

All the CCCHs are the GSM logical channels (RACH, AGCH, and PCH). In this case, all the necessary GPRS information is transmitted on SI 13 (SI 15 messages are no longer used for this). If PBCCH is used in a cell, a single PDCH carries PBCCH, but this PDCH may have several PBCCH blocks (between one and four). Each block is made up of four consecutive time slots (PDCH carried by four consecutive TDMA frames).

The number of PBCCH blocks existing in the cell is given by the parameter BS_PBCCH_BLKS, which is broadcast on the PSI of the first PBCCH block (block 0). If many PCCCHs are declared for the cell (this number is provided by the parameter BS_PCC_CHANS transmitted on SI 13), only one of them carries PBCCH (between one and four PBCCH blocks in total, on the same PDCH).

Figure 6-7
The PBCH in GPRS.

PBCCH: Packet System Information
(PSI 1, 2, 3bis, 4, 5, and 13)

BTS

PBCCH is indicated on SI 13 on BCCH.

PBCCH, if used, carries the PSI 1, 2, 3bis, 4, 5, and 13. PBCCH, if used, is not necessarily transmitted on the beacon frequency of the cell, and even not necessarily transmitted on the same TDMA frame as the one carrying BCCH. Frequency hopping may occur on the PBCCH; therefore, the frequency description for PBCCH channel is transmitted on the BCCH (SI 13). Note that in GPRS Phase I, PBCCH is not used and all the relevant GPRS-related information is carried by BCCH.

System Information Type 13 (SI 13)

SI 13 message is broadcast by the network on the BCCH. The message provides the mobile station with GPRS cell-specific, access-related information. The information in this message should be the same as provided in the PSI 13 message on PACCH. If GPRS is required, the MS reads the SI 13 message. SI 13 may indicate if PBCCH is present in the cell.

If PBCCH is present in the cell, the MS camps on it. If PBCCH is not present in the cell, the necessary system information related to GPRS is contained in the SI 13 message (and extended to the SI 14 and SI 15 messages if necessary).

SI 13 Message Contains One of the Two Indications

If PBCCH is present,

- Channel description for the PBCCH
  - **TN** Time slot number used for PBCCH and PCCCHs
  - **TSC** Training sequence code for PBCCH and corresponding PCCCH
  - **ARFCN** Nonhopping radio frequency absolute RF channel number
- Localization of PSI type 1 information

  If PBCCH is not present,

- The *routing area code* (RAC)

- Options available in GPRS cell include
  - *Network Mode of Operation* (NMO) (mode I, mode II, mode III)
  - ACCESS_BURST_TYPE (PRACH on 8 or 11 bits)
- Network control order parameters (NC0, NC1, NC2)
- GPRS power control parameters

Network Control

It will be possible for the network to order the mobile stations to send measurement reports to the network and to suspend its normal cell reselection and accept decisions from the network instead. The degree to which the mobile station resigns its radio network control is variable and is ordered in detail by the parameter NETWORK_CONTROL ORDER.

The following actions are possible to order to the mobile stations, as illustrated in Figure 6-8:

- **NC0 (normal mobile station control)**   The mobile station performs autonomous cell reselection.

- **NC1 (mobile station control with measurement reports)**   The mobile station sends measurement reports to the network according to additional information in the message NC1. It continues its normal cell reselection.

- **NC2 (network control)**   The mobile station sends measurement reports to the network according to additional information in the message NC2. It does not perform cell reselection on its own, and can only make a cell reselection according to a cell reselection command received from the network.

Two parameters are broadcast on the PBCCH and are valid in packet transfer and packet idle modes respectively for all mobile stations in the cell:

- **NETWORK_CONTROL ORDER**   Can also be sent individually to a mobile station on PACCH, in which case it overrides the broadcast parameter

- **REPORTING_PERIOD**   The interval of time between the measurements

Figure 6-8
Network control orders.

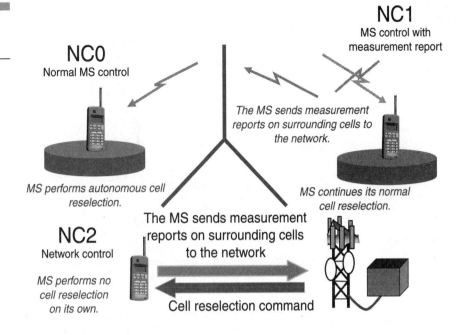

How the Mobile Knows the PDCH Configuration

The mobile station has to decode some parameters broadcast by the network, in order to know the packet channel multiframing and how to access the network. The following list examines the decoding scheme, which is also shown in Figure 6-9:

■ **PBCCH description** The mobile station first decodes the BCCH channel of the initially selected cell. One BCCH is always transmitted on SI 13 (if the cell supports the GPRS service), which indicates all the GPRS relevant parameters, among others, if specific Packet Common Control Channels are used in the cell for GPRS (PBCCH, PRACH, PAGCH, and PPCH). If that is the case, a PCCCH description is sent on SI 13 (on BCCH), indicating that the PDCH is carrying PBCCH. Otherwise, if no PCCCH description is sent on SI 13, all the necessary GPRS information is transmitted on BCCH SI 13 and the mobile station decodes the BCCH of the neighboring cells for the cell reselection process. If PBCCH is used, it carries all the necessary

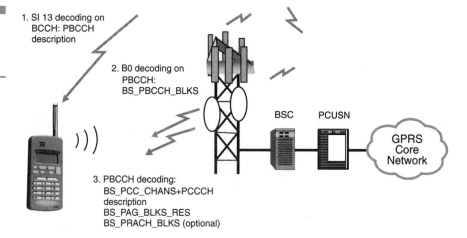

Figure 6-9
The mobile station
learns by decoding
the information.

information about the neighboring cells for the cell reselection process, and the mobile station only has to perform level measurements on the neighboring cells declared in the GPRS_BA_list (sent on BCCH or on PBCCH if it exists).

- **PBCCH first block (B0) decoding** If a PBCCH description is sent on BCCH SI 13, the mobile station listens to the first PBCCH block (B0) on the PDCH carrying PBCCH. The information contained in this first block indicates how many PBCCH blocks are used (and necessarily transmitted on the first blocks of this time slot, according to the multiframe blocks order) through the parameter BS_PBCCH_BLKS.

- **Packet System Information decoding** The mobile station then decodes all the PBCCH blocks, carrying the PSI 1, 2, 3bis, 4, 5, and 13. All the necessary information for the mobile station is transmitted on the PSI, and some redundancy occurs with the SI broadcast on BCCH so that the mobile station only needs to decode PBCCH. The following parameters, among others, are transmitted on the PSI (PBCCH):

  - **BS_PCC_CHANS** Indicates the number of PDCH carrying PCCCH channels

  - **BS PAG_BLKS_RES** Indicates the number of blocks on which paging (PPCH) is forbidden on each PDCH carrying PCCCH channels

  - **BS_PRACH BLKS (optional)** Indicates the number of blocks (UL) reserved for random access (initial resource request on PRACH)

- **BS_PCC_CHANS and BS_PAG BLKS_RES**   Useful for the mobile station to determine the paging subgroup it will use when in DRX mode (after expiration of the timer: NON DRX TIMER)

Example of PBCCH + PCCCH Configuration

The multiplexing of the packet channels on a PDCH is not fixed like in the GSM system. In GPRS, it is managed by some parameters and the following block order: B0, B6, B3, B9, B1, B7, B4, B10, B2, B8, B5, and B11.

Figure 6-10 shows the parameters that manage the multiplexing of the packet channels on the same PDCH:

- **BS_PBCCH_BLKS**   This indicates the number of PBCCH blocks used to broadcast the PSI.

These blocks are always the first blocks according to the multiframe blocks order. The parameter is broadcast on the first PBCCH block (B0). If many PCCCH channels (the total number of PCCCH channels is given by the parameter BS_PCC_CHANS broadcast on PBCCH) are present, PDTCH may be transmitted at the PBCCH blocks position in the other PCCCH channels (because PBCCH blocks are only transmitted on one single time slot in the cell).

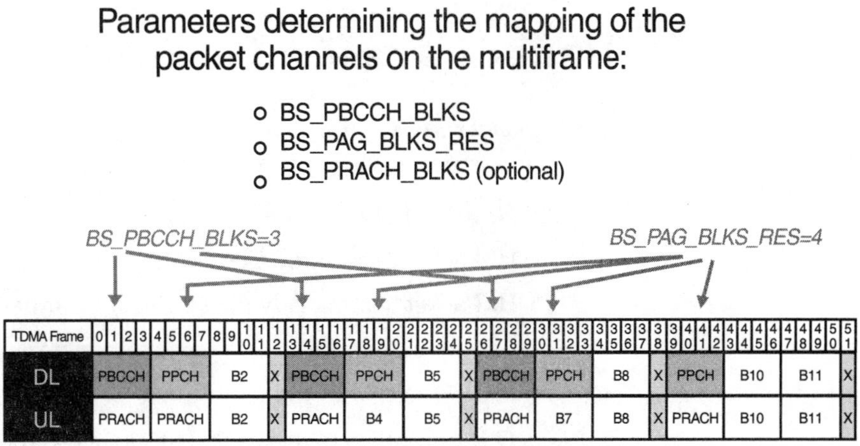

Figure 6-10
An example of the packet logical channel configuration.

Parameters determining the mapping of the packet channels on the multiframe:

o BS_PBCCH_BLKS
o BS_PAG_BLKS_RES
o BS_PRACH_BLKS (optional)

■ **BS_PAG_BLKS_RES (number of blocks for access grant)**
Indicates the number of PCCCH blocks following the PBCCH blocks
(according to the multiframe blocks order) on which paging is
forbidden.

These blocks are thus reserved for PAGCH, and eventually
PDTCH/PACCH/PTCCH. If one of these blocks is used for PAGCH,
then the corresponding UL channels can be used for network access
(PRACH), so *uplink state flag* (USF) is annotated as being free in the
MAC header of the radio block. If one of these blocks is used for
PDTCH (DL), then the corresponding UL block can carry PRACH (USF
is annotated as being free on the corresponding DL radio block)
channels only if the mobile station was not requested to send a PACCH
on this PDCH UL (for ACK, for example).

■ **BS_PRACH_BLKS (optional)**   Indicates the number of blocks
reserved for network access (PRACH channels are reserved by blocks of
four time slots even if an access burst requires only one time slot to be
transmitted because all the other packet channels need four time slots
to transmit their information on the RLC/MAC radio block), starting at
B0 and according to the multiframe block order.

The other PRACH blocks (four PRACH time slots) are indicated
through an USF that is annotated as being free in the MAC header of
the corresponding DL radio block. The nonreserved blocks may carry
either PAGCH or PDTCH/PACCH/PTCCH (unlike the GSM system
where a CCCH never carries traffic).

Packet Traffic Channels

The Packet Traffic Channels are the only ones used in GPRS Phase I. Figure 6-11 provides an example of the Packet Traffic Channel configurations.

PDTCH: Packet Data Traffic Channel

PACCH: Packet Associated Control Channel

PTCCH: Packet Timing Advance Control Channel

The main difference between the PDTCH and the TCH logical channels
in GSM consists of two points:

■ PDTCH is allocated either for downlink or for uplink, but not for both
(simplex channel).

Figure 6-11
Packet Traffic
Channels.

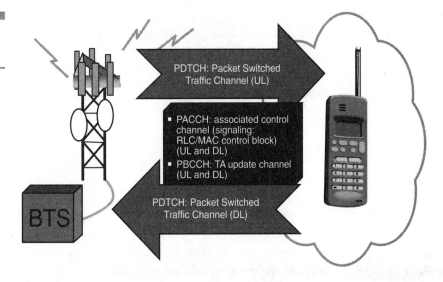

■ The control signaling (PACCH) associated with the assigned PDTCH
is described only by its PDCH, but has no fixed position on this PDCH
(radio blocks). As a conclusion, the mobile station may transmit a
PACCH whenever UL, even by preempting the block position of a
PDTCH, and the mobile station has to continuously monitor the DL of
the DL control time slot indicated in the Packet Resource or
Immediate Assignment.

Up to eight PDTCHs may be allocated to one subscriber (on the same
TDMA) and up to eight mobile stations may share the same PDCH. Currently, the actual numbers are no more than three total time slots (two
downlinks and one uplink for most manufacturers). Some manufacturers'
mobile stations can support four downlink and one uplink time slots (for
example, Motorola). Many of the network resources (aside from the mobile
stations) are limited to the two downlink and one uplink time slots due to
timing advance situations regardless of the capability of the mobile station.
The MAC layer manages this allocation and sharing, assigning the different blocks of the same PDCH to different users through static or dynamic
allocation. Figure 6-12 shows an example of the allocation.

In GPRS Phase I, the mobile station will be able to manage only up to
three time slots simultaneously, and only the static allocation (bitmap) will
be used.

The PACCH carries all the dedicated signaling (it corresponds to
SDCCH + SACCH + FACCH). It always carries RLC/MAC control blocks

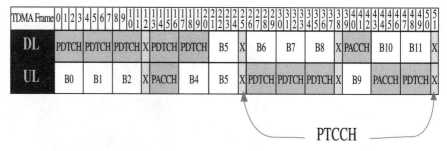

Uplink PDTCH and downlink PDTCH on the same
time slot are disassociated.

(RLC/MAC layer signaling). It is used for paging, for PSI 1 and PSI 13 broadcast in packet transfer mode, ACK messages (answer to polling), RR request, assignment, and reassignment.

On the uplink, the PTCCH carries an access burst in order for the BTS to update the TA value, and on the downlink, it carries the updated value of the TA to apply on the mobile station (even in the case of DL packet transfer because of the UL acknowledgments necessary).

The PACCH and PDTCH positions (PDCH number) are provided to the mobile station in the Immediate Assignment or in the Resource Assignment message. It corresponds to one of the PDCHs assigned for PDTCH.

One-Phase and Two-Phase Access

The mobile station initiates a packet transfer by sending a Packet Channel Request message on PRACH (or RACH). In short access or one-phase access, this message contains all the information needed for the channel establishment:

- Number of requested blocks (in short access case only)
- Radio priorities (11-bit message only)

The BSS acknowledges the request by sending a Packet Immediate Assignment message carried on the PAGCH (or AGCH) and containing the description of the physical channels (PDCHs) reserved for the mobile station. Two types of access are available: one-phase and two-phase access. These are illustrated in Figure 6-13.

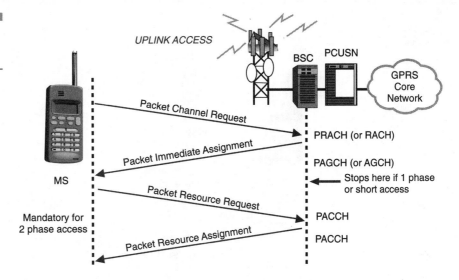

Figure 6-13
One-phase and two-phase access.

The one-phase access is somewhat insecure and requires an efficient resolution mechanism; it will be introduced later. Only the two-phase access is accepted by the network in GPRS Phase I (even if the mobile station may require one-phase access).

The two-phase access can be initiated by the mobile station or by the network. In this case, the mobile station receives a single block on the PAGCH (or AGCH) and responds with a Packet Resource Request message on the PACCH (RLC/MAC control block sent on the allocated UL block) containing the complete description of the requested resources for the uplink transfer. The Packet Resource Request response has two ways of allocating specific blocks on PDCHs: either static or dynamic allocation. Static allocation sends a bitmap to the mobile station that indicates the reserved blocks for uplink transmission in time. Dynamic allocation assigns an USF to each mobile station, so that they will be able to recognize when they are able to transmit on the UL. Hence, the mobile station may start data transmission on the allocated PDCH.

Packet Uplink Assignment

Several steps are involved when the mobile station needs access to enter packet transfer mode, as shown in Figure 6-14. One such example is the

Figure 6-14
Packet Uplink
Assignment message.

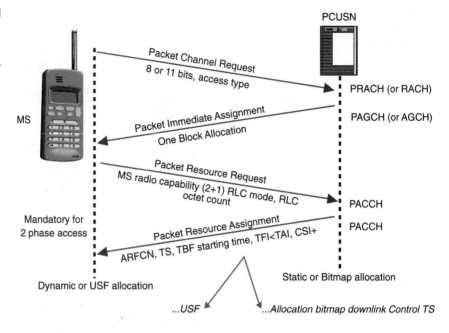

need to have an uplink time slot assigned so that the mobile station can begin transferring data. The sequence uses the following dialogue:

Access Request

In order to result in a GPRS Attach Request, the mobile station sends a Packet Channel Request on the PRACH, which contains the following information: access type (one- or two-phase access) and some reference bits (random bits, priority bits). The Packet Channel Request could be found in two formats: 8 or 11 bits.

Access Answer

The network answers with a Packet Immediate Assignment message (which permits the allocation of a PACCH resource) sent on any PAGCH block on the same PCCCH. This message contains the single allocated block where the mobile station can make its request.

Resource Request

When the mobile has received a Packet Immediate Assignment message, it can perform its request with a Packet Resource Request sent on the PACCH that has been allocated. This message contains the mobile station radio capability (2 + 1 or 3 + 1, and so on), the number of octets for user data (RLC octet count), the LLC-PDU type, the *Temporary Logical Link Identity* (TLLI), and the RLC mode (acknowledged or unacknowledged).

Resource Assignment

The network will respond by sending Radio Resource Assignment messages on one or more PDCHs to be used by the mobile station for the *Temporary Block Flow* (TBF) in a Packet Uplink Assignment message on PACCH. This message contains the *absolute radio frequency channel number* (ARFCN) parameter, one or several time slots, TBF starting time, *Temporary Block Flow Identifier* (TBFI), TA, channel coding scheme (CS1, CS2, CS3, CS4), TLLI for contention resolution, and

- If using static allocation, it contains bitmap allocation and downlink control time slot.
- If using dynamic allocation, it contains USF.

For each TBF (UL or DL), the PCU assigns specific blocks on one or several time slots, allocated for the TBF. These blocks are described in a bitmap that is transmitted to the mobile station or not transmitted according to the case (UL or DL packet transfer, static or dynamic UL resource allocation, and so on). In GPRS, a single mobile station can be assigned up to eight PDCHs for one packet transfer or up to eight mobile stations may simultaneously share the same PDCH.

When several mobile stations share the same PDCH, the multiplexing of these mobile stations is managed by the PCU through a bitmap indicating which mobile station should use each block. The uplink and downlink may have different bitmaps in GPRS, because the GPRS channels are asymmetric and independent. However, the bitmaps of the mobile stations sharing the same way as on the same PDCH are disjointed (similar to the Radio Site Masks of several BTSs sharing the same *Pulse Coded Modulation* [PCM] links on the Abis interface) in order to avoid collisions.

For an UL packet transfer, each mobile station knows when to transmit because of the bitmap transmission in the Packet Resource Assignment (or in the Immediate Assignment for short or one-phase access) if static allocation is used, or from the USF transmitted on the successive DL blocks (MAC header) if dynamic allocation is used.

For a DL packet transfer, the bitmap is never transmitted (it remains in the PCU). Therefore, the mobile stations sharing the same PDCH DL have to listen to all the DL blocks transmitted on this time slot, and they know which mobile station the block is destined to by decoding the *Temporary Flow Identifier* (TFI), transmitted in the RLC header of each RLC data block, and initially assigned to the mobile station in the Packet Resource Assignment (or Immediate Assignment).

Static Uplink (UL) Allocation

In UL static allocation, a mobile station that has requested packet resources for UL network access is allocated in the following ways:

- It is allocated from one to eight time slots (PDCH) on the same TDMA. Nortel Networks' implementation, for example, makes sure that the assigned time slots are always successive. The first mobile stations (GPRS Phase II) will be 3 + 1 terminal equipment (three time slots downlink plus one time slot uplink).

- It is allocated by a TBF STARTING TIME (optional) indicating the position of the first block to use as a TDMA number (for all the allocated time slots for this mobile station). This TDMA number is the *Frame Number* ([FN] on 22 bits) modulo three minutes. It is coded onto 2 bytes. The TBF STARTING TIME description is similar to the STARTING TIME description in some GSM messages. If the TBF STARTING TIME is not used, the mobile station applies the bitmap on the next received blocks of all the allocated time slots.

- It is allocated by a DL control time slot, indicating which of the assigned PDCHs on the downlink channel will the mobile station continuously monitor in order to listen for broadcast control messages for communication supervision (primarily acknowledgment messages and resource reassignments).

- It is allocated by a bitmap, which indicates the specific blocks dedicated to the mobile station on each assigned time slot of the TDMA. The bitmap is determined by the PCU (traffic management functionality) and avoids collisions (simultaneous access of several mobile stations on the same PDCH block). A bitmap is always determined inside the PCU, whatever the communication mode, but it is transmitted to the mobile station in the case of static UL allocation only. In this case, the mobile station decodes the bitmap as rectangular (as many fields as TS allocated).

In the previous list and in Figure 6-15, two mobile stations have been allocated time slot 2 on a same TDMA, but collisions will be avoided thanks to the bitmap, which allocates different blocks to the mobile stations on time slot 2.

Temporary Block Flow

Continuing with the previous example, the focus remains only on time slot 2 (UL and DL), allocated to both mobile stations (mobile station 1 and mobile station 2), which is shown in Figure 6-16.

■ On the downlink, only mobile station 1 continuously (all blocks) monitors time slot 2 because this time slot has been allocated to mobile station 1 as the DL CONTROL time slot in the Packet

Figure 6-15
Static uplink
allocation.

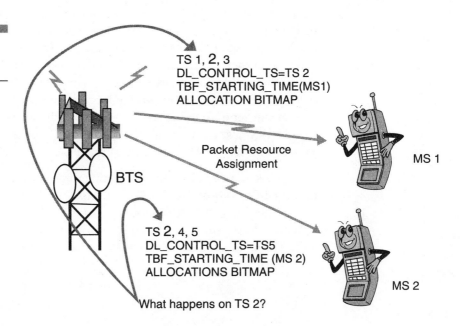

TS 1, 2, 3
DL_CONTROL_TS=TS 2
TBF_STARTING_TIME(MS1)
ALLOCATION BITMAP

Packet Resource
Assignment

BTS

MS 1

TS 2, 4, 5
DL_CONTROL_TS=TS5
TBF_STARTING_TIME (MS 2)
ALLOCATIONS BITMAP

MS 2

What happens on TS 2?

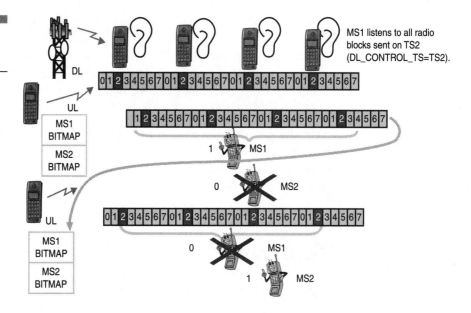

Figure 6-16
Temporary Block
Flow.

Resource Assignment (or Immediate Assignment in the case of short or one-phase access), whereas the DL control time slot of mobile station 2 is time slot 5 (which is not considered in this example).

■ On the uplink, mobile station 1 and mobile station 2 are multiplexed through their bitmaps for their respective access. They successively transmit or not on time slot 2 UL, according to the successive values decoded on their bitmaps. When the bitmap shows a 0 concerning time slot 2, the mobile station cannot access time slot 2 UL on the corresponding block, but when the bitmap shows a 1, the mobile station can send information on time slot 2 UL, either with an RLC data block (most often) on a PDTCH logical channel or an RLC/MAC control block on a PACCH logical channel.

The uplink and downlink access starting times (the same for UL and DL) of each mobile station are indicated by the parameter TBF STARTING TIME transmitted with the other parameters in the Packet Resource Assignment. In the previous example, we assume that the TBF STARTING TIME is the same for both mobile stations (to simplify the drawing).

Dynamic Uplink (UL) Allocation

In dynamic UL allocation, shown in Figure 6-17, the bitmap is not transmitted for uplink access, but a number, called an USF and coded onto 3 bits, is attributed to each mobile station. The access is dynamically granted to each mobile station communicating on the same PDCH UL through the DL transmission of this USF in a traffic radio block (or in a dummy radio block if no DL traffic is present, where only relevant information consists in the USF). Note that one USF is assigned for each time slot allocated, so as many USF are present as time slots allocated (they may be equal because they concern different time slots). A TBF STARTING TIME may also be used (optional), and has the same role as for the static allocation case.

Temporary Block Flow (TBF) for Dynamic Allocation

Continuing with the case of dynamic allocation for UL network access, each mobile station decodes the USF transmitted on all blocks of the allocated

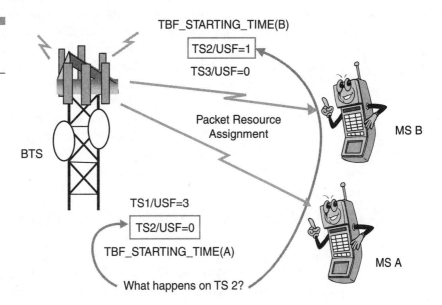

Figure 6-17
Dynamic uplink allocation.

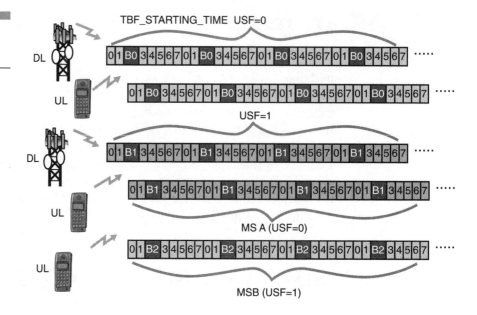

Figure 6-18
TBF for dynamic allocation.

time slot DL, as shown in Figure 6-18. The focus is still on time slot 2, which has been allocated for both mobile station A and mobile station B.

When a mobile station decodes its USF on one radio block DL, it is allowed to transmit on the next radio block UL. This way, the PCU enables or disables the access on the successive blocks of the same PDCH, for mobile stations that have been allocated a common PDCH. (This way, the PCU also manages the transmission on polling answer by mobile station UL, by coordinating the *Relative Reserved Block Period* [RRBP] field, indicating for a polled mobile station the block position where the answer is expected on the network side and the USF transmission on the previous DL radio block for this mobile station.)

RLC/MAC Block Structure

A radio block (also called an RLC/MAC block) consists of one MAC header, one RLC data block or one RLC/MAC control block, and one *Block Check Sequence* (BCS). Some fields are specific to the uplink or to the downlink way. Figure 6-19 shows the data blocks side by side.

Figure 6-19
The RLC/MAC
data blocks.

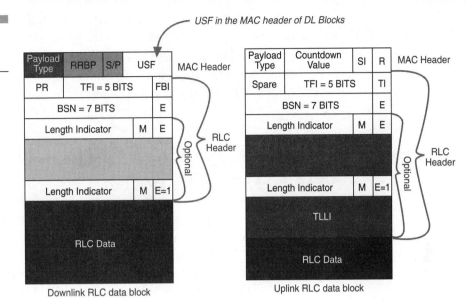

Downlink RLC data block

Uplink RLC data block

MAC header contains the following:

- **Uplink state flag (USF—3 bits)** This is used to identify users for UL transmission, or to characterize a PRACH.

- **Type (2 bits)** The payload type that identifies the type of block that follows (RLC data block or RLC/MAC control block).

- **Polling control (3 bits)** One *Supplementary/Polling* (S/P) bit to poll the mobile station (so that it sends an acknowledgment message) and two RRBP bits to tell the mobile station where to send the acknowledgment message.

The RLC data block's header contains the following:

- *Block Sequence Number* **(BSN—7 bits)** This carries the absolute BSN modulo 128 of each RLC data block within the TBF.

- **Temporary Flow Identifier (TFI—5 bits)** The TFI identifies the TBF to which the RLC block belongs.

- *Power Reduction* **(PR—2 bits)** This indicates the power level reduction of the next RLC blocks, which is based on GSM power control.

- *Final Block Identifier* **(FBI—1 bit)** The FBI indicates that the RLC data block is the last one of the downlink TBF (DL).

- **Length indicators (optional bytes)** Length delimits the LLC frames when an RLC block contains more than one LLC frame.
- **Temporary Logical Link Identifier (TLLI—several bytes)** This identifier identifies the logical link established between the user and the *Serving GPRS Support Node* (SGSN) (UL only).

Temporary Block Flow—Uplink (UL) Data Transfer

This procedure, shown in Figure 6-20, is an example of message sequence for the uplink data transfer with one resource reallocation and possible RLC data blocks retransmissions (we assume that the transfer mode is ACK). A contention resolution mechanism is adopted in order to avoid two mobile stations perceiving the same UL channel as their own. If static allocation is used, this mechanism is the UL bitmap transmission for the allocated UL PDCH, and if dynamic allocation is used, the USF field transmitted on the DL radio blocks dynamically indicates the attribution of the next UL radio block.

Figure 6-20
The uplink data transfer.

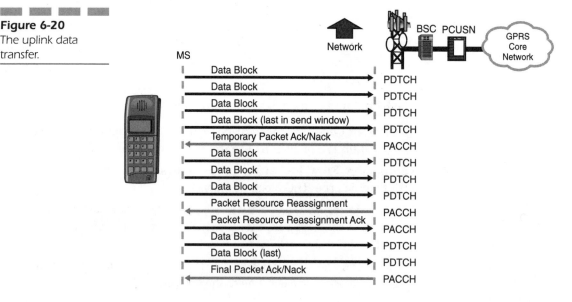

Two modes of transmission are available: acknowledged and unacknowledged. The mode of transmission is indicated in the PDP context activation (*quality of service* [QoS] field).

In the acknowledged mode of DL transmission, the mobile station is regularly polled through the S/P bit in the MAC header, and it should transmit an acknowledgment bitmap in the correct block indicated in the RRBP field of the MAC header (2 bits). This bitmap enables the network to selectively retransmit received blocks with errors.

In the acknowledged mode of UL transmission, the network regularly sends temporary acknowledgments to the mobile station.

In any case, the sending window is 64 blocks (UL and DL). This window is shifted after each temporary or final acknowledgment message. All acknowledgment messages are transmitted on a PACCH (RLC/MAC control block).

Downlink (DL) Resource Allocation

Downlink resources are allocated to the mobile station via the Packet Downlink Assignment message. This message will detail all the time slots that the mobile station may receive data on for a particular transaction. Each complete data transfer is allocated a TBF known by the identifier as already discussed (the TFI). The TFI is part of each uplink/downlink RLC data block and consists of 7 bits in the uplink and 5 bits on the downlink. The TFI for a specific mobile station is also specified in the Packet Downlink Assignment message. The mobile station has to receive and decode all the RLC/MAC blocks on its allocated time slots to ascertain if the TFI contained in the block is the appropriately assigned TFI. When the mobile station identifies a block with its allocated TFI, it will decode and process the data block. This is shown in Figure 6-21.

The network initiates packet transfer to a mobile station in standby state by sending a Packet Paging Request message in the downlink PCCH (or PCH).

- The mobile station responds by requesting a channel.
- The Packet Paging Response message contains the TLLI as well as a complete LLC frame, including the TLLI.
- The mobility management state of the mobile station then becomes ready state.

Figure 6-21
Downlink data
access.

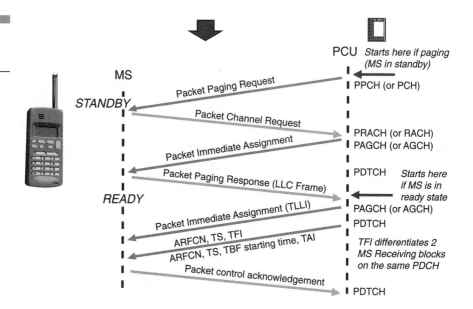

The network initiates transmission of a packet to a mobile station in the ready state by using the Packet Downlink Assignment message on PAGCH (or AGCH). This message includes the list of PDCHs that will be used for a downloading transfer; if available, timing advance and power control information are also included. The network sends the radio blocks belonging to one Temporary Block Flow on the assigned downlink channels, as shown in Figure 6-22.

Timing Advance Updating Procedure

The timing advance procedure is necessary because a proper value for timing advance has to be used for the uplink transmission (radio data or control blocks). Figure 6-23 provides a representation of the timing advance procedure. The mobile station's initial timing advance is calculated on the access burst the same as GSM. The estimated timing advance value is passed to the mobile station via the Packet Immediate Assignment message. The mobile station uses the value until continuous timing advance provides a new value.

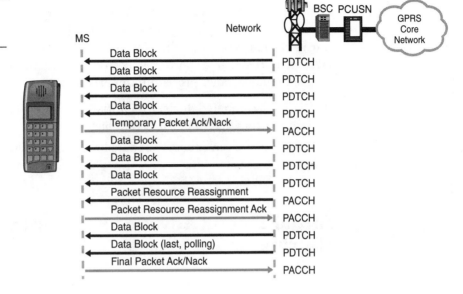

Figure 6-22
Downlink data transfer.

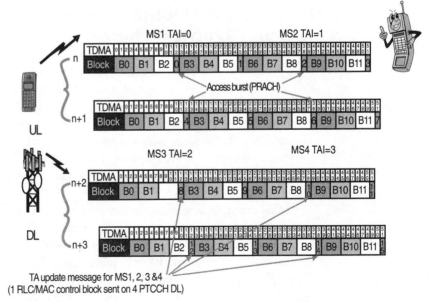

Figure 6-23
TA update procedure.

In continuous timing advance, the mobile sends a special access burst in an idle slot for the network to derive the necessary timing. In the downlink, the network sends a timing advance via the PACCH, which is transmitted during the idle time slots of the 52 multiframes. The *Timing Advance Index* (TAI) gives the mobile station the position to send the access burst.

The timing advance procedure comprises two parts:

- The TA is estimated during the initial phase of the data transfer (initial estimation from the access burst).

- The TA is also continuously updated while the mobile station remains in transfer mode.

Initial Timing Advance Estimation (During Access Phase)

This is made from the access burst carrying the PRACH. This first estimation is sent to the mobile station through the first available message (Packet Immediate Assignment or Packet Resource Assignment).

Continuous Update Procedure (During Transfer State)

This procedure is carried on one of the allocated PDCH using PRACH UL and PACCH DL. Each mobile station is assigned in the Packet Resource Assignment (for UL or DL packer transfer) a specific PDCH among those allocated and an idle slot (indicated by the TAI) on this PDCH to be used for TA updating. The mobile station sends on its assigned slot a special access burst and receives the TA value on the subsequent TA message. The TAI is coded on 4 bits, creating 16 different positions in groups of 8, 52-time-slot multiframes.

In this example, a mobile station transmits an access burst on idle slot 0 of multiframe (n), and receives the TA value on TA message in multiframes (n + 2) and (n + 3). If this message is not correct, then the mobile station listens to the next three TA messages that also contain the updated TA value. Therefore, the updated TA value for up to four mobile stations' that sent an access burst on PTCCH of the multiframes n and n + 1 is contained in multiframes n + 2 to n + 7.

Identifiers Limitations

Several different identifiers were used during the development of the GPRS. Many retain some or all of their values from GSM. Others are specific to GPRS. Regardless of the origin, the limitations concern the maximum number of mobile stations that the MAC layer can multiplex onto the same time slot (PDCH).

The Temporary Flow Identifier

The TFI differentiates two TBFs, which have a common PDCH (uplink or downlink) allocated. For any direction of the packet transfer, the opposite direction may have some acknowledgments, so collisions may occur for communications on the same time slot even though the packet transfer directions are different. In other words, the TFI differentiates the TBFs corresponding to different mobile stations that have been allocated the same PDCH. This may be the case to control only the TBFs of the same TDMA.

The TFI is coded on 5 bits, so 32 (2^5) different TFI values are possible. Consequently, the TFI limitation is, at most, 32 mobile stations: TDMA 32 UL + 32 DL.

The Timing Advance Index

The TAI differentiates two mobile stations communicating on the same time slot for the TA continuous update procedure. The mobile stations communicating on the same time slot may transfer or receive data packets because mobile stations receiving data packets also need to update their TA for acknowledgment transmission.

The TAI is coded with four bits, so 16 (2^4) different TAI values are possible. Consequently, the TAI limitation is at most 16 mobile stations/time slots. Based on the USF limitation and future improvements (dynamic UL allocation), some manufacturers have split this limitation into a maximum of eight mobile stations on the uplink and eight mobile stations on the downlink using the same PDCH.

The Uplink State Flag

The USF gives successive UL access on a given PDCH to different mobile stations that have been allocated this same time slot (PDCH), when dynamic allocation is used for UL channel assignment. DL channel assignment has no specific limitation except the nonoverlapping between the DL bitmaps of different mobile stations communicating on the same PDCH (this bitmap remaining in the PCU).

The USF is coded on 3 bits, so 8 (2^3) different USF values are possible. Consequently, if dynamic allocation is used for UL network access, the USF limitation is eight mobile stations per time slot on the UL.

RLC/MAC Block

Network Layer Protocol Data Units (NL-PDU) are transmitted over the air interface by using the Logical Link Control (LLC) and the RLC/MAC protocols. The *Subnetwork-Dependent Convergence Protocol* (SNDPC) transforms packets into LLC frames. LLC frames (currently variable up to a maximum of 1,600 octets) are then segmented into RLC data blocks (or RLC/MAC control blocks), which are formatted by the physical layer into blocks of four successive time slots on the same physical channel (one per frame), as shown in Figure 6-24. The rate of RLC/MAC data blocks is one block every 20 ms.

Figure 6-24
The RLC/MAC blocks.

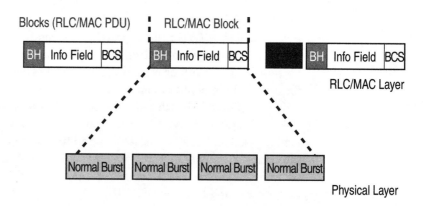

BCS = Block Check Sequence   BH = Block Header

Activity at the BSS

Data and signaling messages arrive at the BSS via the G_b interface and through the Network Service/Frame Relay layer. The frames arriving at the PCU pass through the *BSS GPRS Protocol* (BSSGP) where the information and signaling messages are separated into LLC frames, *GPRS mobility management* (GMM) information, and *network management* (NM) information. For data and signaling messages destined to the GPRS mobile station, the LLC frames pass through a relay entity (LLC relay) before being positioned into the RLC and MAC, respectively. The RLC/MAC layer provides services for information transfer over the physical layer of the GPRS interface. These functions include *backward error correction* (BEC) procedures enabled by selective retransmission of blocks of data with errors. The MAC function arbitrates access to the shared medium between multiple mobile stations and the GPRS network.

Medium Access Control (MAC) Layer

The MAC layer provides capability for multiple mobile stations to share a common transmission medium. It interfaces directly with the physical layer. For the uplink (such as a mobile attempting to originate access), the MAC layer plays the role of an arbitrator managing the limited physical resources among many competing requestors. The reservation protocol used for contention resolution among the various mobile station devices is based on the Slotted Aloha protocol. In addition, many services may be competing for the same limited radio resource within a single mobile station. The MAC layer coordination function is responsible for resolving these contentions.

For the downlink (such as a mobile termination), the MAC layer aids in the queuing and scheduling of the access attempts. Contention resolution is not required for the downlink because only a single transmitter is present.

The MAC layer also prioritizes the data to be sent. Signaling data is given a higher priority than user data. Signaling as well as user data is multiplexed onto the transmission medium. The MAC layer enables multiple mobiles to share a common transmission medium. The transmission medium may be a single physical channel or multiple physical channels. In the TDMA world, a physical channel is simply a single TDMA time slot.

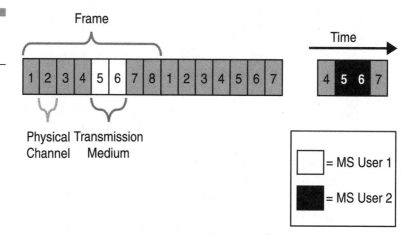

Figure 6-25
Transmissions at the MAC layer.

When multiple physical channels are allocated for the transmission medium, the mobile station essentially sends data in parallel. This provides the capability to increase the data rate between the mobile station and the network. In Figure 6-25, two mobile users are actively transmitting streams of data. Time slots 5 and 6 are allocated as the shared transmission medium. The MAC layer on the network side schedules the mobiles in order to stagger the transmission of data. So, user 1 may send data on the one TDMA frame and user 2 may send data on the next TDMA frame. The standards enable the network to schedule a maximum of eight mobiles to use the same transmission medium.

Key Identifiers for the MAC Layer

The MAC layer uses several identifiers to transfer data. A brief description of two of them is listed in the following section.

- **Temporary Block Flow**   Used to identify a series of RLC/MAC blocks to/from a specific mobile station. The TBF is unique for a given direction (uplink/downlink). Each mobile station occupying a radio resource is assigned a TBF for the duration of the data transfer. Because data transfers are typically bursts of data followed by idle time, the TBF is temporary; it only lasts until all RLC/MAC blocks have been transferred and acknowledged.

- **Temporary Flow Identity**   Uniquely identifies each TBF for a given direction. The TBF, TFI, and direction uniquely identify a RLC data block. The message type together with the TBF, TFI, and the direction designates the RLC/MAC control message.

The following concepts are depicted in Figure 6-26.

- **Data Burst 1** The MAC layer in the mobile station receives an LLC frame that is ready for transfer. The mobile station communicates with the network and ultimately receives a TFI that will be used to identify all consecutive (that is, one data burst) RLC blocks that are transferred. The MAC layer then segments the LLC frame and encapsulates it with an RLC header containing the TFI. Once all RLC blocks have been transferred and acknowledged (in the ACK mode), the TFI is released. At this point, the radio resources are not required; the TBF no longer exists.

- **Idle** The mobile station has no data to transfer even though GPRS data services remain active.

- **Data Burst 2** The mobile station has additional data to transfer. It notifies the network in order to establish another TBF. The TFI corresponding to this TBF will most likely be different from that corresponding to the first TBF. Again, once all RLC blocks have been transferred and acknowledged, the TFI is released and the TBF disappears.

This procedure can happen repeatedly until the GPRS service is complete.

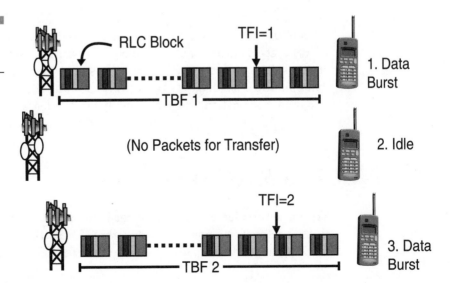

Figure 6-26
Identifiers at the
MAC Layer.

Channel Allocation and the MAC Layer

The uplink state flag enables the network to control the multiplexing of mobile stations. The USF field is included in the header of each RLC/MAC data block sent on the downlink. It designates the mobile that can transmit data in that particular time slot of the next uplink radio block. Eight possible USF values can be assigned, so a maximum of eight users can be multiplexed in the same time slot.

Three allocation modes are possible in the MAC layer: fixed allocation, dynamic allocation, and extended dynamic allocation. Each of these modes applies to the uplink transfer from the mobile station to the network.

- In fixed allocation mode, the network communicates to the base station a set of physical charnels that the mobile is to use for the transfer of data. The network can specify either a consecutive or nonconsecutive range of physical channels using an allocation bitmap.

- In dynamic allocation mode, the mobile reads the USF from the header of each RLC/MAC data block. When the mobile station detects its assigned USF, it can transmit either a single RLC/MAC block or a set of four RLC/MAC blocks. Because the mobile station is constantly monitoring the USF, the allocation scheme can change dynamically.

- Extended dynamic allocation works much in the same way as the dynamic allocation. The main difference is that the system may specify a range of physical channels for the mobile to transmit data on. This provides a higher throughput in the uplink direction.

In each case, the network controls the mode that is to be used. Not all GPRS systems are capable of supporting each of these modes. All mobile stations that support GPRS services support the fixed allocation and dynamic allocation modes. The network may either support fixed allocation or dynamic allocation. The extended dynamic allocation mode is optional.

The MAC Header

The main function of the MAC layer is the control of multiple mobile stations sharing a common resource on the GPRS air interface. The RLC data block is passed down to the MAC layer where a MAC header is added. The format of the MAC header is dependent upon the direction of data transfer. The fields in the header are shown in Figure 6-27 and lay out as follows in Table 6-1.

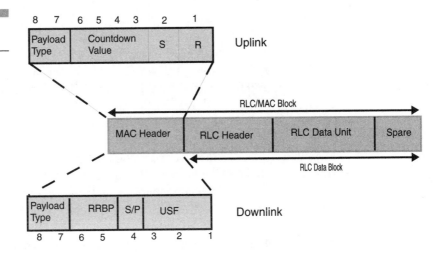

Figure 6-27
The MAC header.

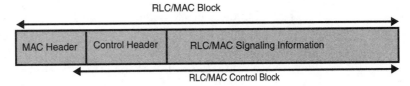

Table 6-1 Radio Link Control (RLC) Layer		
USF	Uplink state flag is used to indicate which mobile station is allocated the GPRS resource.
S/P	Supplementary/Polling bit is used to indicate whether the RRBP field is active.
RRBP	Relative Reserved Block Period is used to specify that a single uplink block is being used as a Packet Associated Control Channel (PACCH).
Payload type	Payload type defines the type of information in the payload area: either data or signaling information.
SI	Stall Indicator is used to signal whether the transmission has stalled.
Countdown value	This is sent by the mobile station (on the uplink) to the network so that the network can calculate the number of radio blocks remaining in the current uplink allocation of resources.
R	Retry bit indicates whether the mobile station was transmitted on the channel.

Radio Link Control (RLC) Layer

The RLC layer is primarily responsible for segmenting and reassembling data sent over the air interface. The frames used in the LLC layer are much too big to send over the air. Thus, the RLC layer segments or breaks the LLC frame into blocks, and encapsulates each block forming an RLC block. A BSN designates each RLC block. The BSN is contained in a field of the block header. Upon receipt of an RLC block, the RLC layer reverses the action required to send the data. First, the BSN is used to arrange the RLC block in sequential order. Then the header is stripped off the block and the blocks are reassembled into LLC frames.

This layer supports two modes of operation: acknowledged and unacknowledged. The acknowledged mode enables selective retransmission. In this mode, the BSN is also used to request the retransmission of a missing or undelivered block. The unacknowledged mode of operation does not guarantee the arrival of the transmitted RLC blocks. This mode is important to applications that require a constant delay. The RLC layer increases the reliability of the air interface by providing BEC, which enables selective transmission.

The RLC data block consists of the RLC header, RLC data field, and spare bits. Each RLC data block may be encoded using any of the available coding schemes (CS-1, CS-2, CS-3, or CS-4) and will affect the degree of segmentation and reassembly. If the contents of an LLC-PDU do not fill an entire RLC block, the beginning of the next LLC-PDU will be used to fill up the remaining positions. However, if the LLC-PDU was the last in the current transmission block, the RLC data block will be filled with spare bits (padding). The structure of the RLC data blocks is dependent upon the transmission direction (uplink or downlink).

Mobile-Originated Access Message Sequence

Several messages are sent between the mobile station and the network in order to establish a connection path. Figure 6-28 is an example of the mobile-originated sequence messages.

1. The mobile station that is not currently transmitting packets on the uplink receives an LLC frame from the upper layers. It sends a Packet

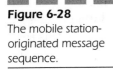

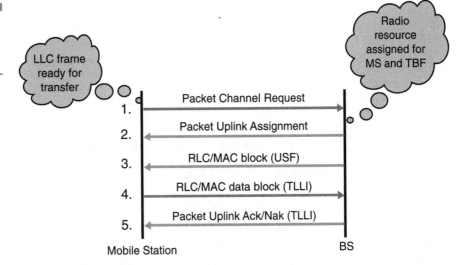

Channel Request on the PRACH to request resources from the BS. For GPRS, this message is either 8 or 11 bits long. This small size provides fast access. The Packet Channel Request message may specify the throughput, RLC mode, ACK/NAK, and the priority of the message.

2. If the BS has resources available for the mobile station, the network reserves a radio resource (or time slot) for the mobile station and assigns a TFI. The network communicates the assignment to the mobile station with the Packet Uplink Assignment message. This message also contains the mode of operation (fixed allocation or dynamic allocation). In the fixed allocation mode, the assignment message will contain a list of assigned slots for the mobile station to use.

3. In the dynamic allocation mode, the mobile station reads the USF from the incoming RLC/MAC data blocks. The USF specifies the physical channel on which the mobile station is to transfer data.

4. The mobile station starts transmitting data using the RLC/MAC data block.

5. For acknowledged service, the BS sends an acknowledgment over the PACCH on the downlink. RLC is a sliding window algorithm, so if a valid ACK is not received, the mobile station selectively retransmits the unacknowledged packets.

Such a mechanism enables the mobile station to quickly go back and forth between idle mode and packet data transfer mode; hence, radio resources are utilized only when data is available to transmit. If the BS does not have resources available to process the mobile station request, it may choose to queue the request until resources are available. In this case, the BS sends a Packet Queue Notification message back to the mobile station on the PCCCH.

The Radio Resource (RR) State Model

In the beginning of Chapter 5, "Main GPRS Procedures," the models of how and when the resources are allocated were discussed. However, to reiterate the point of grabbing a resource, when the mobile station is activated for data services, it will be in one of two states, as shown in Figure 6-29.

- Packet transfer mode
- Packet idle mode

In the packet transfer mode, the mobile station is allocated an RR providing a TBF for a physical point-to-point connection on one or more physical channels. This enables the unidirectional transfer of the LLC frames between the network and the mobile station. In the packet idle mode of operation, the radio resource providing the TBF does not exist. The mobile station monitors the relevant paging subchannels on PCCCH. The transition from packet idle to packet transfer mode can be triggered implicitly

Figure 6-29
The states for radio resource allocation.

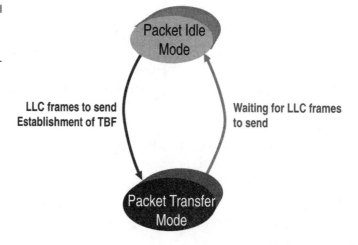

whenever a higher layer needs to transfer an LLC frame. This accommodates packet data, which is typically characterized by discontinuous traffic with short bursts of high activity interleaved with periods of idle time.

BSS GPRS Protocol (BSSGP) Layer

The BSSGP provides a connectionless link with unconfirmed data transfer between the BSSGP and the SGSN. The BSSGP shown as the model in Figure 6-30 has a reliable link below it in the form of the network service. The network service is a Frame Relay network. The Base Station Subsystem GPRS Protocol (BSSGP) acts as an interface between the LLC frames and the RLC/MAC blocks in the BSS and as an interface between the RLC/MAC derived information and the LLC frames in the SGSN.

Also, a BSS could receive information from the underlying Frame Relay network in multiple ways. Basically, the BSS controls several mobiles and each mobile could get information on different *Network Service-Virtual Links* (NS-VLs). The BSSGP also provides radio-related QoS and routing information to facilitate data transfer between the BSS and the SGSN.

BSSGP uses a *BSSGP Virtual Connection Identifier* (BVCI) for information transfer between the SGSN and the BSS. The connection over the Gb, interface is called a *BSSGP Virtual Connection* (BVC). The BVC is made up of the BVCI and the *Network Service Entity Identifier* (NSEI). In the BSS,

Figure 6-30
The BSSGP layer in relation to the other layers.

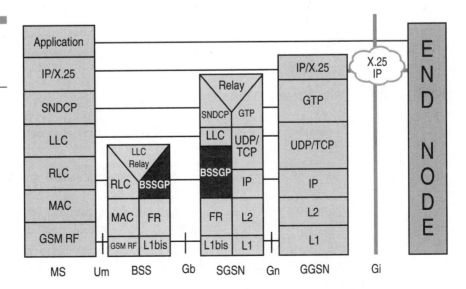

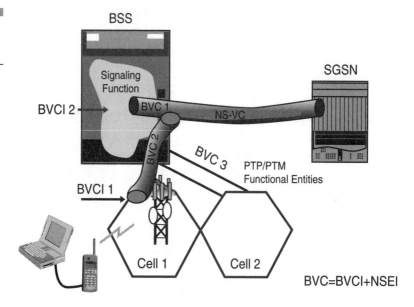

a BVCI is allocated for each cell that supports GPRS. For each new cell introduced in the BSS area, a new BVCI context is allocated. In the SGSN, the BVCI context consists of at least one queue for LLC-PDUs.

In a BSS, a BVC is connected to a cell or to a functional entity such as a signaling entity. A signaling functional entity is used for the functions like paging. One or more signaling entities exist per BSS. The BVCs are statically provisioned. Figure 6-31 shows the BSSGP with the provisioning of multiple BVCIs.

Channel Coding

Channel coding in *basic GSM operation* is performed using two codes: a block code and a convolutional code.

- The block code corresponds to the block code defined in the GSM Recommendations 05.03. The block code receives an input block of 240 bits and adds four zero tail bits at the end of the input block. The output of the block code is consequently a block of 244 bits.

- A convolutional code adds redundancy bits in order to protect the information. A convolutional encoder contains memory. This property differentiates a convolutional code from a block code. A convolutional code can be defined by three variables: n, k, and K. The value n corresponds to the number of bits at the output of the encoder, k to the number of bits at the input of the block, and K to the memory of the encoder. The ratio, r, of the code is defined as $r = k / n$.

Let's consider a convolutional code with the following values: k is equal to 1, n to 2, and K to 5. This convolutional code uses a rate of $r = 1/2$ and a delay of $K = 5$, which means that it will add a redundant bit for each input bit. The convolutional code uses 5 consecutive bits in order to compute the redundancy bit. As the convolutional code is a half-rate convolutional code, a block of 488 bits is generated. These 488 bits are punctured in order to produce a block of 456 bits.

The block of 456 bits produced by the convolutional code is then passed to the interleaver. Before applying the channel coding, the 260 bits of a GSM speech frame are divided in three different classes according to their function and importance. The most important class is the class Ia containing 50 bits. The class Ib is next in importance, which contains 132 bits. The class II is the least important, which contains the remaining 78 bits. The different classes are coded differently. First of all, the class Ia bits are block coded. Three parity bits, used for error detection, are added to the 50 class Ia bits. The resultant 53 bits are added to the class Ib bits. Four zero bits are added to this block of 185 bits (50 + 3 + 132). A convolutional code, with $r = \frac{1}{2}$ and $K = 5$, is then applied, obtaining an output block of 378 bits. The class II bits are added, without any protection, to the output block of the convolutional coder. An output block of 456 bits is finally obtained.

An interleaver rearranges a group of bits in a particular way. It is used in combination with FEC codes in order to improve the performance of the error correction mechanisms. The interleaving decreases the possibility of losing whole bursts during the transmission, by dispersing the errors. Because the errors less concentrated, it is then easier to correct them. This is shown in Figure 6-32.

A burst in GSM transmits 2 blocks of 57 data bits each. Therefore, the 456 bits corresponding to the output of the channel coder fit into four bursts ($4 \times 114 = 456$). The 456 bits are divided into 8 blocks of 57 bits. The first block of 57 bits contains the bit numbers (0, 8, 16, . . . 448), the second one contains the bit numbers (1, 9, 17, . . . 449), and so on. The last block of 57 bits contains the bit numbers (7, 15, . . . 455). The first 4 blocks of 57 bits are placed in the even-numbered bits of 4 bursts. The other 4 blocks of 57 bits are placed in the odd-numbered bits of the same 4 bursts. Therefore, the interleaving depth of the GSM interleaving for control channels is four and

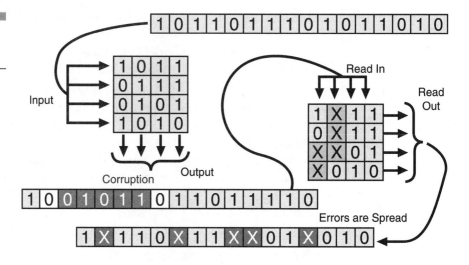

Figure 6-32
Interleaving spreads the errors.

a new data block starts every four bursts. The interleaver for control channels is called a block rectangular interleaver.

Interleaving Speech Channels

The block of 456 bits, obtained after the channel coding, is then divided in eight blocks of 57 bits in the same way as it is explained in the previous paragraph. However, these eight blocks of 57 bits are distributed differently. The first four blocks of 57 bits are placed in the even-numbered bits of four consecutive bursts. The other four blocks of 57 bits are placed in the odd-numbered bits of the next four bursts. The interleaving depth of the GSM interleaving for speech channels is then eight. A new data block also starts every four bursts. The interleaver for speech channels is called a block diagonal interleaver.

Interleaving For the GSM Data TCH Channels

A particular interleaving scheme, with an interleaving depth equal to 22, is applied to the block of 456 bits obtained after the channel coding. The block is divided into 16 blocks of 24 bits each, 2 blocks of 18 bits each, 2 blocks of

12 bits each, and 2 blocks of 6 bits each. It is spread over 22 bursts in the following way:

1. The first and the twenty-second bursts carry 1 block of 6 bits each.
2. The second and the twenty-first bursts carry 1 block of 12 bits each.
3. The third and the twentieth bursts carry 1 block of 18 bits each.
4. From the fourth to the nineteenth burst, 1 block of 24 bits is placed in each burst.

A burst will then carry information from five or six consecutive data blocks. The data blocks are interleaved diagonally. A new data block starts every four bursts.

Channel Coding in GPRS

Four different coding schemes, CS-1 to CS-4, are defined for the radio blocks carrying RLC data blocks. These are summarized in Figure 6-33 with the coding scheme, the throughput in data rates, and the coding ratio. For the radio blocks carrying RLC/MAC control blocks, code CS-1, similar to SDCCH coding in GSM, is always used. Two specific coding schemes are

Figure 6-33
The four coding schemes used.

4 Coding Schemes:
CS-1: same coding as SDCCH in GSM
CS-2 and CS-3=CS-1+ punctured bits
CS-4: no coding for error correction

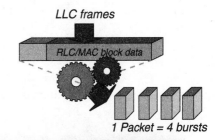

Coding Scheme	Code Rate	RLC/MAC Block data size (bytes)	RLC/MAC max throughput (Kbps)
CS-1	1/2	20	8
CS-2	2/3	30	12
CS-3	3/4	36	14.4
CS-4	1	50	20

used for the access burst: one with 8 information bits and one with 11 information bits. For implementation reasons, it is convenient to regard the error detection as part of the physical link layer (GSM RF layer), even though the backward error correction procedure belongs to the RLC layer.

For CS-2 and CS-3, the second step consists in precoding the USF for a specific protection of this information. For CS-1 to CS-3, adding four tail bits (at 0) and applying a punctured convolutional encoding of rate 1/2 (for error detection and correction on the receiver side) give the desired coding rate (22.8 Kbps). Indeed, a radio block, resulting in 456 bits after channel coding, is delivered every 20 ms.

For CS-4, no convolutional coding is applied for error correction. The following list covers the differences to help compare the performance of the different coding schemes:

- CS-1 gives the largest throughput for low C/I[1] values (because the other CSs do not protect the data sufficiently, so the RLC layer requires more RLC data blocks retransmissions).

- CS-2 and CS-3 show similar performances.

- CS-4 is the most adapted for high C/I values.

By dynamically adapting the coding scheme to the radio channel conditions, it is possible to optimize the communication performances. CS-1 is mandatory for the BSS, whereas CS-1 to CS-4 are mandatory for mobile stations. Only CS-1 and CS-2 are used in GRPS Phase II.

The CCU, equivalent to the *Signal Processing Unit* (SPU), is inside the BTS. It is in charge of the channel coding. Initially, the coding scheme is determined on a TDMA-per-TDMA basis at the OMC, so it will be fixed in time. In GPRS Phase II, the most adapted coding scheme is determined dynamically by the PCU according to the quality measured on the radio link during the communication.

---

[1] C/I is the channel-to-interference ratio used in GSM. Channel interference can be either adjacent channel interference or cochannel interference based on these standards. Regardless of the interference source the point is that the more unreliable the data link (air) the less reliable throughput. Thus, coding is used to protect the data blocks at the expense of overhead ratios being higher.

Coding Scheme-1 (CS-1)

For CS-1, shown in Figure 6-34, 40 bits are used for BCS to increase protection. It has 3 USF bits, 181 header and data bits, and 40 BCS bits, totaling 224 bits. After adding four tails bits (always set to 0), the half-rate convolutional coding for forward error correction (the same as for TCH and SDCCH in GSM) is applied. USF are decoded as part of the data. The payload rate (radio block except BCS and USF) is 9.05 Kbps (181 bits/20 ms) and the data maximum throughput accepted by this coding scheme on one time slot is 8 Kbps.

Coding Scheme-2 (CS-2)

CS-2 and CS-3 are punctured versions of CS-1. For CS-2, shown in Figure 6-35, 16 bits are used for BCS using a *Cyclic Redundancy Check* (CRC) code. USF bits are precoded (6 bits in total) to increase robustness. The half-rate convolutional coding (the same as for CS-1) is applied (to the precoded USF + headers [MAC and RLC] + data + BCS) and 132 bits puncture the result in order to get the 456 desired bits.

The USF bits (first 12 bits) are not affected by puncturing. USF can be decoded either as a block code or as part of the data. The payload (radio block except USF and BCS) rate is 13.4 Kbps (268 bits/20 ms) and the

Figure 6-34
Coding scheme-1.

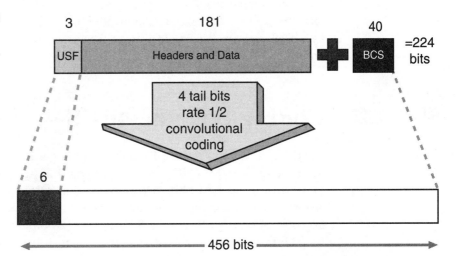

Figure 6-35
Coding scheme-2.

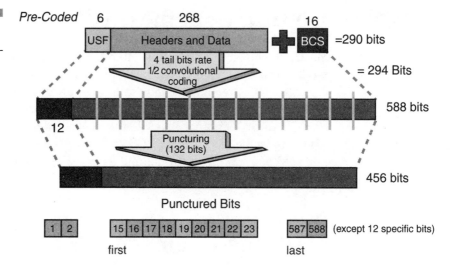

maximum data throughput accepted by this coding scheme is 12 Kbps per time slot.

Coding Scheme-3 (CS-3)

For CS-3, shown in Figure 6-36, 16 bits are used for BCS and a CRC code is applied. USF bits are precoded (6 bits in total) to increase robustness. The half-rate convolutional coding (the same for CS-1) is punctured. The USF bits (first 12 bits) are not affected by puncturing. USF can be decoded either as a block code or as part of the data. The payload (radio block except USF and BCS) rate is 15.6 Kbps (312 bits/20 ms).

Coding Scheme-4 (CS-4)

For CS-4, shown in Figure 6-37, no FEC coding is applied. Sixteen bits are used for BCS and a CRC code is applied. USF are precoded (12 bits in total) with the same 12-bit code as for CS-2 and CS-3. USF can be decoded either as a block code or as part of the data. The payload (radio block except USF and BCS) rate is 21.4 Kbps (428 bits/20 ms) and the maximum data throughput accepted by this coding scheme is 20 Kbps per time slot.

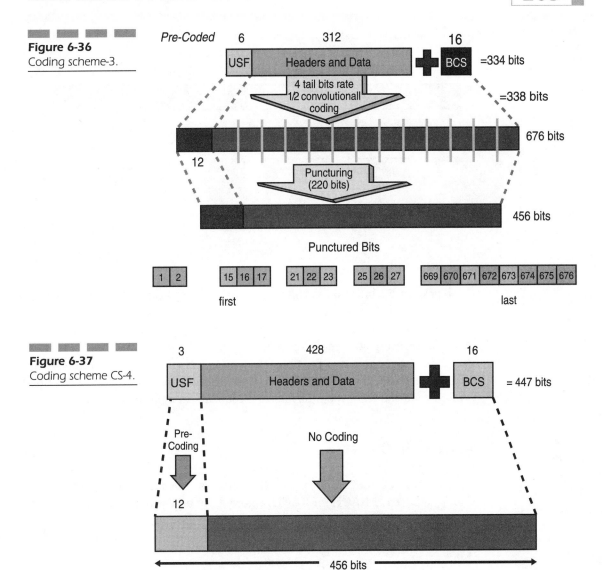

Figure 6-36
Coding scheme-3.

Figure 6-37
Coding scheme CS-4.

Normal Burst

The normal burst, shown in Figure 6-38 is used on the PDTCH, PACCH, PTCCH, PAGCH, PPCH, PBCCH, and PNCH. The information relative to these packet channels is transmitted onto radio blocks mapped onto four consecutive normal bursts. A normal burst contains 26 bits for the training sequence, 2 blocks of 58 bits for information. More precisely, it contains

Figure 6-38
Normal burst format.

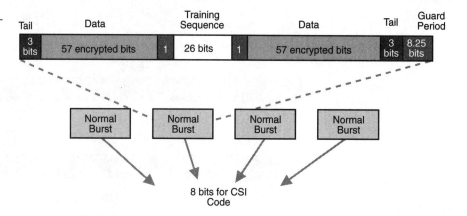

Normal Burst: used for PDTCH, PACCH, PTCCH, PAGCH, PPCH and PBCCH

2 blocks of 57 information bits and two stealing flags, which are used in GPRS to indicate the coding scheme used (CS-1 to CS-4). It also contains two sets of three tail bits, and 8.25 bits for the guard period.

The training sequence depends on the BCC (BSIC) of the cell, so eight training sequences are possible, showing a very low correlation between each other in order to correctly differentiate the information destined to one mobile station from the information broadcast in neighboring cells.

Access Burst

The access burst format is used for network access and TA update (UL) only, using the PRACH (or RACH) channel. An access burst contains 41 bits for the training sequence, 36 bits for the information, and 8 and 3 tail bits at the beginning and the end of the burst, respectively. The guard period has 68.25 bits.

RLC Layer Segmentation

LLC frames are variable length, whereas RLC/MAC blocks are fixed length (depending though on the coding scheme used; the coding scheme is directly linked to the protection added to the rough information before transmission on the radio interface). Therefore, one LLC frame may be spread onto several RLC/MAC blocks, as shown in Figure 6-39, and conversely, one

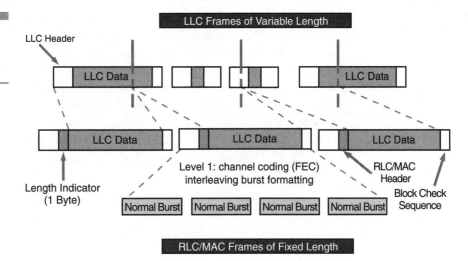

Figure 6-39
The RLC/MAC data blocks segment the LLC data.

RLC/MAC block may carry several LLC frames (not necessarily completely). This is why the RLC header has optional-length indicators in order to delimitate the different LLC frames information.

Quality of Service (QoS)

A QoS profile is associated with each PDP context and is considered a single parameter with multiple data transfer attributes. These are defined as

- Precedence class
- Delay class
- Reliability class
- Peak throughput class
- Mean throughput class

Several possible QoS profiles are defined by these various attributes and as such a PLMN may support a limited number of the profiles. This is implementation-dependent. During the QoS profile negotiation, it is possible for the mobile station to request a value for each of the QoS attributes, including the *Home Location Register* (HLR) stored subscribed default values. QoS parameters are normally negotiated at subscription or during call setup. The network will negotiate each attribute at a level that is consistent with available GPRS resources.

The use of the interface and the group of protocols to handle data transfer between the mobile station and the PCUSN are clearly defined in the role of the two communicating devices. The mobile station and the BSS/PCU combination use the air interface to communicate in a secure manner, with reliable data transfer across a hostile medium (that being the air). These combined interfaces and protocols provide more efficient radio resource utilization. From this interface, the next step is for the mobile station to communicate across these interfaces to the SGSN, covered in Chapter 7, "X.25, Internets, Intranets, and Extranets."

X.25, Internets, Intranets, and Extranets

Objectives

Upon completion of this chapter, you should be able to

■ Understand the benefits of circuit-switched vs. packet-switched networks.

■ Discuss the original concept of X.25 packet switching.

■ Understand the services of the Internet and intranet.

■ Describe how the TCP/IP works.

■ Describe how data is transferred across the IP networks.

Modes of Switching

In any given network, switching points are interconnected to a form of mesh through which voice and data calls are routed from one device to another, based on the terminal's address. The addressing methods are somewhat unique to the network being used. The network provides connectivity for every device in a way that is consistent with the infrastructure. As a user dials a call, for example, resources (circuit-switched) are allocated for the duration of the connection. Each network, however, uses a different form of switching, but could be an overlay to a different network. This is the case with many of the data networks today. These data networks are overlays to the circuit-switched telephone network regardless of the traffic they carry. GPRS is similar in that the IP-based GPRS data network is an overlay to the GSM circuit-switched network.

Circuit Switching

Circuit switching is the process of passing calls across a given path (voice, data, video, text, or multimedia) that is created on a temporary duration for the express purpose of carrying the traffic between the two endpoints. After the connection is used, the terminals are disconnected and the circuit is released and made available to another short-duration user for the duration of a call. For this connection, only two (typically) parties use the entire circuit for the duration of the connection (the sender and receiver). The entire circuit is tied up even if no information is being exchanged at the time. The connection is nailed up so that no one else can use the same circuit at the same time. This is viewed as a typical phone call.

On many long-distance networks, when a call setup is necessary, each leg of the call repeats the call setup procedure until the last exchange in the loop is reached. In essence, the call is being built by the signaling as progress is occurring on a link-by-link basis. As each link is added to the connection, the network is building the entire circuit across town or across the country. This method is an inefficient use of the circuitry. Although the call reaches its end destination, several complications could arise. Regardless of the complications, the outcome is the same; the carrier ties up the network and never completes the call. This is no big deal when discussing one call. However, when a network carries hundreds of millions of calls per day, this accumulated lost time is extensive and expensive. Moreover, when dealing with the use of circuit-switched calls, the provider still has to accommodate voluminous demands depending on a demand-dial basis. The inefficient use of the circuitry requires that the carriers overbuild the network to accommodate all forms of traffic demands on user demand. Preplanning is rather futile, as the user population may make one connection today and 100 connections tomorrow.

As described previously, the connection orientation of the call also mandates that the circuit must be built (either physically or logically) and the connection must be established (someone must answer and say hello) before a conversation can take place. In reality, if the connection never goes through, then the provider continues to attempt the connection with limited success. Although this is an inefficient use of the network, it is the way that many networks were built.

Packet Switching Defined

Packet switching is a means of taking a very large file of information (data) and delivering it to a piece of hardware or software. From this interface, the hardware or software breaks the information down into smaller, more manageable pieces. As these pieces are broken down, additional overhead is applied to the original segment of data. This overhead is used for control of the information. Because the information is segmented, the packet service inserts the telephone number of the addressee, along with the segmentation number (packet #1, packet #2, packet #3, and so on), so that the data can be reassembled at the receiving end. Once the overhead is attached to the segmented data (now called a packet), the packet is transmitted across a physical link to a switching system that reads the address information (telephone number) and routes the packet accordingly. This establishes a virtual connection to the distant end, and each packet is sent along the same route as the first packet. The system uses a connection-oriented transport based on a virtual circuit.

The Packet Concept

If an organization has large amounts of data to send, then the data can be delivered to a *packet assembler/disassembler* (PAD). The PAD can be a software package or a piece of hardware outboard of the computer system. The PAD acts as the originating mail clerk in that the originating PAD receives the data and breaks it down into manageable pieces or packets. In the data communications arena, a packet can be a variable length of information, usually up to 128 bytes of data (one page in the example). Other implementers of X.25 services have created packets up to 512 bytes, but the average is 128, as shown in Figure 7-1. The 128-byte capability is also referred to as a *fast select*. The packet-switching system can immediately route the packet to a distant end and pass data of up to 128 bytes (1,024 bits).

Overhead

The PAD then applies some overhead to the packet as follows:

- An opening flag that is made up of 8 bits of information. Using a standard *high-level data link control* (HDLC) framing format, the

Figure 7-1
The X.25 packet.

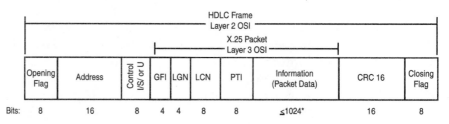

						HDLC Frame Layer 2 OSI			
						X.25 Packet Layer 3 OSI			
Opening Flag	Address	Control I/S or U	GFI	LGN	LCN	PTI	Information (Packet Data)	CRC 16	Closing Flag
Bits: 8	16	8	4	4	8	8	≤1024*	16	8

* Variations can exist using 4,096 bits of data

I = Information
S = Supervisory
U = Unnumbered

opening flag is a sequence of 8 bits that should not be construed as real data.

■ A 16-bit address sequence that is a binary description of the endpoints (the *from* address).

Control information consists of 8 bits of data describing the type of HDLC frame that is traversing the network (this is a notation on the envelope that describes the information inside). These can be supervisory, unnumbered, or information fields. To be more specific, these break down as follows:

■ *Information* **(I)**   Used to transfer data across the link at a rate determined by the receiver and with error detection and correction

■ *Supervisory* **(S)**   Used to determine that the ready state of the devices—*receiver is ready* (RR), *receiver is not ready* (RNR), or *reject* (REJ)

■ *Unnumbered* **(U)**   Used to dictate parameters, such as set modes, disconnect, and so on

Packet-specific information follows the HDLC information. The packet information consists of the following information (this information is similar to the *to* address and the designation of the routing that will be used, such as first class, book rate, and so on):

■ *General format identifier* **(GFI)**   Four bits of information that describe how the data in the packet is being used: from/to an end user, from/to a device controlling the end-user device, and so on.

■ *Logical channel group number* **(LGN)**   Four bits that describe the grouping of channels. Because only 4 bits are available, only eight combinations are used.

■ *Logical channel number* **(LCN)**   An 8-bit description of the actual channel being used. The theoretical number of channels (ports) available is 2,048. Although the number of logical channels can be 2,048, most organizations implement significantly fewer ports or channels.

The next overhead elements consist of the *Packet Type Identifier* (PTI), an 8-bit sequence that describes the type of packet being sent across the network. Six different packet types are used in an X.25-switching network. These packet types define what is expected of the devices across the network.

The variable data field is now inserted. This is where the 128 bytes of information are contained in the packet. The 128-byte field is the standard implementation, but as mentioned, it can be larger (as much as 512 bytes).

Following the information field is the *Cyclic Redundancy Check* (CRC), a 16-bit sequence that will be used for error detection and/or correction. Using a CRC-16, the error detection capability is approximately 99.99 percent. This is to ensure the integrity of the data; rather than having to deliver information over and over again, the concept is to deliver reliable data to the far end.

The closure of the packet is the end-of-frame flag. In the HDLC frame format, this denotes the end of the frame so the switches and related equipment know that nothing follows. The switches then calculate all of the error detection and accept or reject the packet on the basis of the data integrity.

The Packet Network

Figure 7-2 shows a typical network layout. The cloud in the center of the drawing represents the network provided by one of the carriers. From our discussions in earlier chapters, the obvious network configuration is that each of the designated carriers has its own cloud. Actually, each cloud inter-

Figure 7-2
The X.25 network.

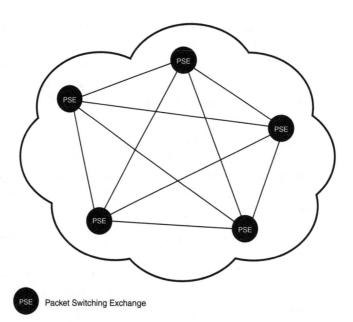

PSE　Packet Switching Exchange

connects to the clouds provided by other carriers so that transparent communications can take place. Once the cloud is established, the next step is to provide the packet-switching systems (called *packet-switching exchanges* [PSEs]). These are nothing more than computers that are capable of reading the address and framing information. The PSEs then route the packet to an appropriate outgoing port to the next downstream neighbor. In many cases, these PSEs are connected to several other PSEs. The packet switches can select the outbound route to the next downstream neighbor on the basis of several variables. The selection process can be the circuit used least, the most direct, the most reliable, or some other predefined variable. Once again you can see the magic that takes place inside the cloud. Now that the network cloud and the packet exchanges are in place, the next step is to connect a user.

The User Connection

Users sign up with the carrier of choice and let the carrier worry about the physical connection. As shown in Figure 7-3, the user connection into the cloud is through a dedicated or leased line. The carrier notifies the *local exchange carrier* (LEC) and orders a leased line at the appropriate speed (in the leased line, this can be up to 64 Kbps on digital circuits). The original network connections back in the initial rollout of X.25 services were on analog lines at up to 9.6 Kbps. A modem was provided by the carrier at the customer end, or the customer purchased and provided it. Now that the modem

Figure 7-3
The user connection.

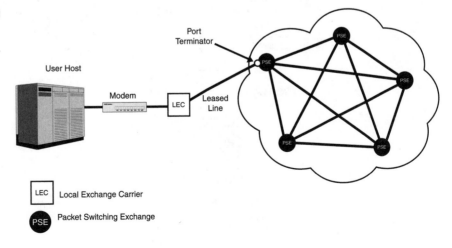

is attached at the customer end, the circuit is terminated in a port on the computer (PSE). This is the incoming port that can be used as a permanent virtual connection or as an incoming-only channel. This is part of the addressing mechanism inside the packet where the PSE reads the packets' originating address. The incoming-only channels are the channel numbers assigned to each customer.

The next step follows the connection where the customer must initiate the PAD function, as shown in Figure 7-4. Remember that the PAD is the hardware or software installed to break the data into smaller pieces, attach the overhead, and forward the packetized data to the network. In the reverse order, the PAD is responsible for receiving the packets, peeling off the overhead, and reassembling the data into a serial data stream to the *Data Terminal Equipment* (DTE), whether it is a terminal or host computer. So the PAD's function is crucial. If a software package is performing the PAD function, then the customer must purchase (or license) the software and install it. If the solution is a piece of hardware, the options are different. The customer might buy the PAD and install it, or the carrier might provide and install it. This hardware can be rented, leased, or sold by the carrier. Now the connection exists on one end of the cloud.

Figure 7-4
The PAD function

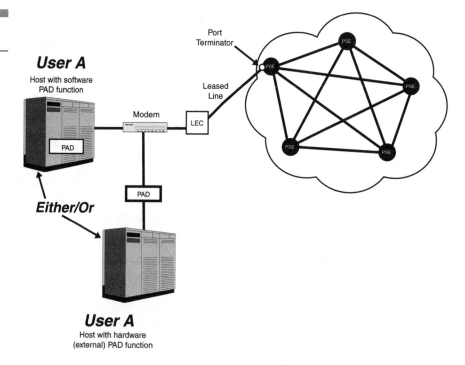

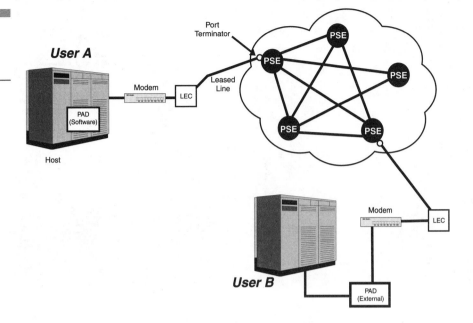

Figure 7-5
Another user is
attached to the
network.

In Figure 7-5, another device is attached on the other end of the cloud. Now the magic of the packet-switching world occurs.

As packets are generated through the network from user device A to B across the network, several things happen, as shown in Figure 7-6.

■ The data is sent serially from the DTE to the PAD (which acts as the *Data Circuit-Terminating Equipment* [DCE] for the computer terminal).

■ The PAD breaks the data down into smaller pieces and add the necessary overhead for delivery.

■ The PAD then routes the packets to the network PSE.

■ The receiving PSE (PSE 1) sees the packet coming in on a logical channel, so it remembers where the packets are coming from. After analyzing the data (performing a CRC on the packet) and verifying that it is all right, the PSE sends back an acknowledgment to the originating device (which is the DCE).

■ The PSE then sends the packet out across an outgoing channel to the next downstream neighbor (PSE 2). This establishes a logical connection between the two devices (PSEs) from the out channel to an in channel at the other end. The logical channel is already there; it is used for the transfer of these specific packets. A virtual connection is

Figure 7-6
The packet sequence.

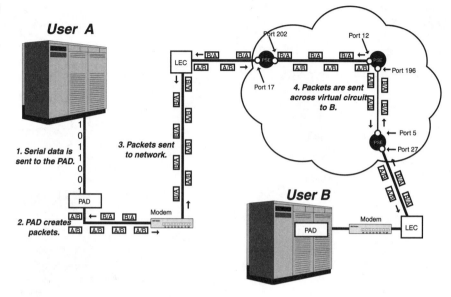

also created, allocating the time slots for the packets from device A to B to run on the virtual circuit.

■ Back at PSE 1, the next packet is sent down from the originating PAD. This is analyzed and acknowledged once again, and then passed along. At the same time this is happening, PSE 2 is sending the first packet to PSE 3. Again, at each step of the way, the packets are opened and a CRC is performed before a packet is actually accepted.

■ At each PSE along the way, the packet is buffered (at PSE 1) until the next receiving PSE accepts the packet and acknowledges it (PSE 2). PSE 1 flushes the packet only after it is acknowledged. Prior to that, the network node (PSE 1) stores it just in case something goes wrong.

■ Don't forget what happens at the receiving end (device B), where the packets arrive in sequential order and are checked and acknowledged. Then the overhead is peeled away so that a serial data stream is delivered to the receiving DTE.

■ This process continues from device A through the network to device B until all data packets are received. Every packet along every step of the link is sent, accepted, acknowledged, and forwarded until all get through. This is the guaranteed delivery of reliable data properly sequenced. The logical link that is established between devices A and B is full duplex. The two devices can be sending and receiving simultaneously.

You can see from the packet delivery process that the packet-switching process is beneficial. However, nothing is perfect. The overhead on the packet, the buffering of multiple packets along the route, the CRC performance at each node along the network, and the final sequenced data delivery all combine to present the risk of serious delays. What you receive in integrity and reliability can be offset in delays across the network. You must always weigh the possibilities and choose the best service.

Benefits of Packets

The real benefit to this method of data delivery ties into the scenario presented in the beginning of this chapter. Remember the problems with the dial-up telephone network and the risk of sending three-quarters of a file transfer only to have a glitch in the transfer? Using the packetized effort, if a glitch occurs, the network might have from 7 to 128 outstanding packets traversing the links. Therefore, instead of scrapping the entire file, the network automatically recovers and resends the packets that were lost or corrupted. This means that the users would save time and money on an error-prone network. Again, the risk of congestion and delay on the network might cause others to look for alternative solutions.

Other Benefits

Beyond these benefits, other benefits can be achieved from the use of the dedicated link into the network. As Figure 7-7 shows, the link is not solely for one user at a time, nor is it for two specific locations. When the organization uses a packet-switching network, the users might need to have multiple simultaneous connections up and running. Therefore, the PAD acts similarly to a *statistical time division multiplexer* (statmux). The statmux capability provides multiple connections into the single device, and samples each of the ports in a sequential mode to determine whether the port has anything to send. If a packet has been prepared, the statmux generates the call request (initiate a call) to the network. The connection is created and the data flows. This assumes that no problems are being experienced on the network.

As a new user logs on and generates a request to send data, the PAD then sets up this connection. Packets are interleaved across the physical

Figure 7-7
Additional users can share the link.

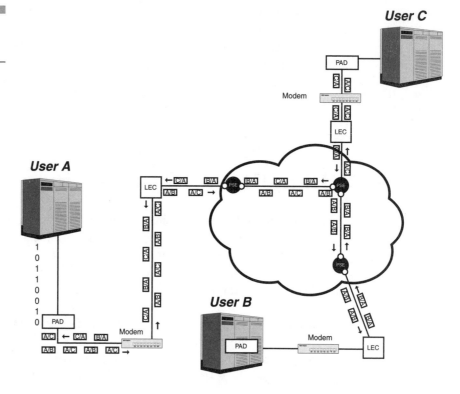

link between or among the various users. One can imagine that not all devices will transmit at the same rate of speed or have the same amount of two-way interactive traffic. Therefore, the statmux function of the PAD interleaves the packets based on an algorithm that enables each device to appear as though a dedicated link is available to it. The use of a statmux is beneficial because it enables users to employ the expensive leased link to the maximum benefit of the organization. This requires fewer physical links and takes advantage of the dead time between transmissions and so on.

Figure 7-8 shows a series of sessions running on a single link, all interleaved. This also shows that packets are not specifically interleaved in the order of A, B, C, and so on. Instead, they can be interleaved on the basis of the flow or delivery method used, such as A, B, A, B, C, B, A, and so on. Herein lies the added benefit of packet switching—the user achieves the data throughput necessary without having a dedicated resource that is only periodically used.

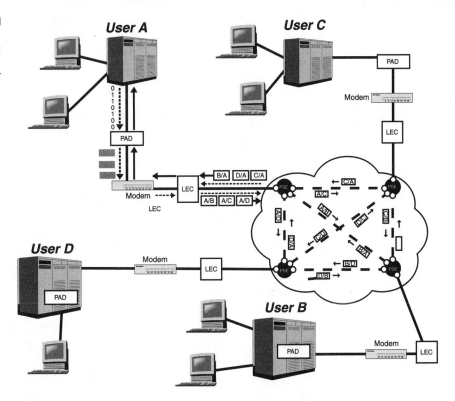

Figure 7-8
Multiple users sharing the same links.

Advantages of Packet Switching

Packet switching is considered by many to be the most efficient means of sharing both public and private network facilities among multiple users. Each packet contains all of the necessary control and routing information to deliver the packet across the network. Packets can be routed independently or as a series that must be maintained and preserved as an entity. The major advantages of using this type of transport system are

- Shared access among multiple users, either in a single company or in multiple organizations.
- Full error control and flow control in place to enable the smooth and efficient transfer of data.
- Transparency of the user data.
- Speed and code conversion capabilities.
- Protection from an intermediate node or link failure.

■ Pricing advantages: because many use the network, the prices are on a per-packet, rather than per-minute, basis.

Other Components of Packet Switching

Although the primary components of packet switching have already been discussed, others are also available. As the evolution of the X.25 standard continued in the early 1970s, several different implementations were enhanced. These include devices or access methods that you should understand. They are summarized so that you can gain an appreciation of how these pieces all work together. Some of the added pieces include

■ The ability to dial into the packet-switching network from an asynchronous modem communication. Although packet switching (a la X.25) is a synchronous transfer system, the need for remote dial-up communications exists. Therefore, the standards bodies included this capability in one of the enhanced versions of the network. As shown in Figure 7-9, a dial-up connection can be made from a user. In this case, the asynchronous communication is a serial data transfer to a network-based PAD. The PAD accepts the serial asynchronous data, collects it, segments it into packets, and establishes the connection to the remote host desired. In this case, the connection is now across the X.25 world, synchronously moving packets across the network. This uses an X.28 protocol to establish the connection between the asynchronous terminal and the PAD.

Figure 7-9
Dial access to X.25 packet switching.

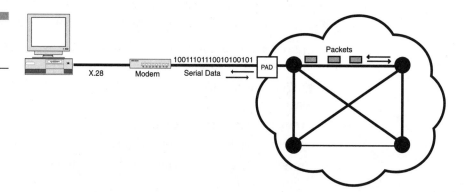

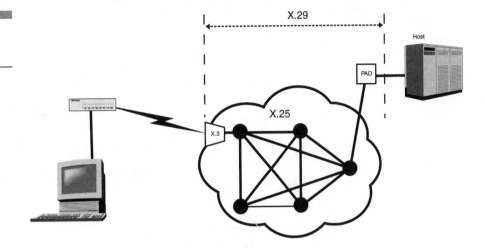

Figure 7-10
Added protocols
for X.25.

- The PAD-to-host arrangement is controlled by the X.29 protocol. Control information is exchanged between a PAD (X.3) and a packet mode DTE or another PAD (X.3). This is shown in Figure 7-10. In this case, as the communication is established between these devices, the X.3 PAD sets the parameters of the remote device. This could be the speed, format, control parameters, or anything that would be appropriate in the file transfer. The X.29 parameters can also provide keyboard conversion into network-usable information.

- The internetworking capability of an X.25 network uses a protocol called X.75. Although this should be user transparent, the network needs the X.75 parameters to provide a gateway between two different packet networks or between networks in different countries. In each case, the gateway function is something that should only concern the network carrier. However, as more organizations install their own private network-switching systems, the need to interconnect to the public data networks rises. Therefore, the internetworking capability is moving closer to the end user's door. Figure 7-11 shows a representation of an X.75 interconnection.

Other Forms of Packet

Another form of packet switching, called a datagram, is used in the industry. A datagram transfer is a form of packet switching. Typically, the

Figure 7-11
X.75
internetworking.

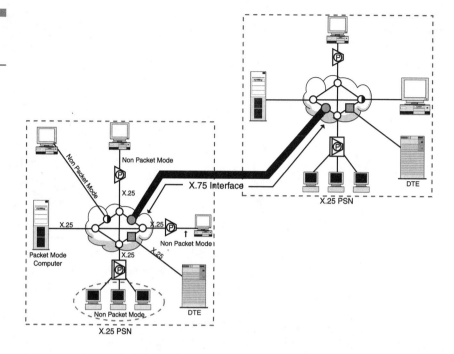

datagram does not follow the same virtual circuit concept. Every packet is sent across the network and can take a different route to get from point A to B. Also, a datagram is not checked at every node along the route; only the header (or address) information is checked for the location of the destination. As each packet is received, it is scanned for the destination address and then immediately routed along to the next node. No CRCs are performed on the data. It is not the network's responsibility to guarantee the integrity of the data; that is left to the receiving device using a higher-level protocol. With this concept of a different path for each packet, the idea of sequentially numbering the packets is lost. The network treats every packet as its own entity (or as packet #1). If a packet gets lost or something else happens, the network does not care. The sequencing and reassembly of the data back into its original form is handled by the receiving end's higher-layer protocols. An example of this is *Transmission Control Protocol / Internet Protocols* (TCP/IP). IP packets are datagrams that are segments of the original data stream. IP also works at Layer 3 of the OSI model. As a network layer protocol, its only concern is to break the data down into packets and make its best effort to deliver the packets. Figure 7-12 shows the OSI model comparison to TCP/IP.

TCP, at the receiving end, is responsible for reassembling the data back into the serial data stream and ensuring the integrity and sequencing of the

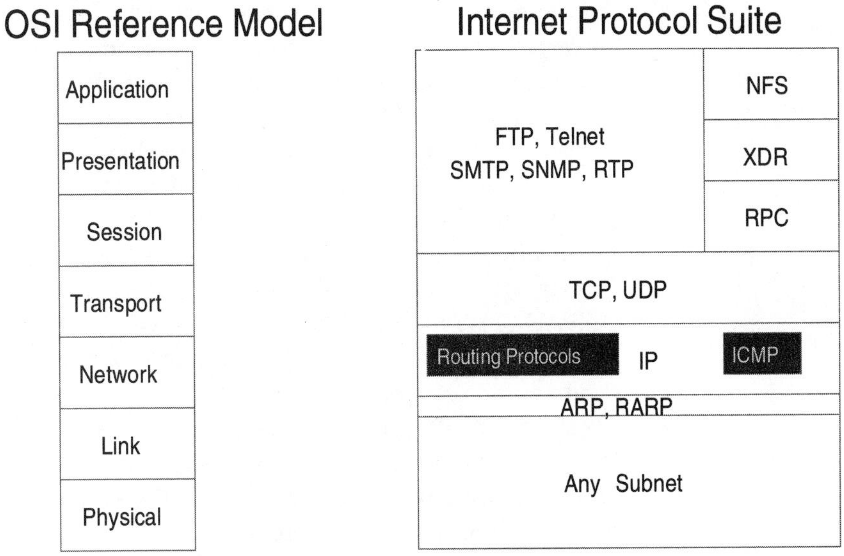

Figure 7-12
OSI and TCP/IP
protocol stacks.

OSI Reference Model

Application
Presentation
Session
Transport
Network
Link
Physical

Internet Protocol Suite

	NFS
FTP, Telnet SMTP, SNMP, RTP	XDR
	RPC

TCP, UDP		
Routing Protocols	IP	ICMP
ARP, RARP		
Any Subnet		

information. TCP is not a network layer protocol; it works at the higher layers of the OSI model (transport and above). TCP/IP is the mainstream of the Internet and has been widely adopted as the protocol stack of choice in many *Local Area Network* (LAN) and *Wide Area Network* (WAN) arenas. Because of its robustness and capability to deal with the packetization of the data, more organizations are using TCP/IP as their WAN protocol. TCP/IP has its problems, but the advantages far outweigh the disadvantages. The industry as a whole and the vendor community in particular has recognized the importance of TCP/IP in the mainstream of products. Just about every LAN *Network Operating System* (NOS) vendor now supports the use of TCP/IP. From a physical network and data link layer, TCP/IP runs on most any topology.

The Internet

The Internet is a network of networks. Most people think it is strictly one network, but nothing is farther from the truth. In reality, the Internet is a huge worldwide communications network linking several smaller regional or national networks together into a homogenous network. This network, composed of smaller ones, enables users to quickly access databases and communicate via *electronic mail* (e-mail). The Internet also connects to thousands of computers and their data. The data is located in databases at universities, schools, research and development labs, government offices, and commercial enterprises. With access to the Internet from anywhere in the world, you can find, receive, and transmit information on virtually any subject. The Internet provides e-mail and hypertext files, full color graphics, audio, and video services. Many agree that the Internet of today is the Information Superhighway of the future.

The original network was developed in the early 1960s by the U.S. government based on the Defense *Advanced Research Project Agency Network* (ARPANET). The U.S. *Department of Defense* (DOD) funded the project in 1969. The purpose of the ARPANET was to provide the existing U.S. *Defense Department Network* (DDN) with a digital highway between its sites and the Pentagon. This high-speed highway was designed to survive a nuclear holocaust or first strike. The data on this network included datagrams (intelligent data packets) to automatically route themselves around a failed segment (link) or nodes on the network.

Protocols and Technologies Enabled the Internet

During the 1970s, the ARPANET gradually transformed into a true Internet as new protocols and technologies became available, and as defense, research, and scientific organizations were added to the network. While these various entities proliferated, the need for robust and secure network partitions (subnetworks) grew. This need led to the development of a specific protocol for transmitting datagrams between the various subnetworks. The creation of TCP/IP protocols facilitated the connection of devices called routers, reliable gateways, and other switching devices. The early routers developed by BBN were designed to fit into the *Digital Equipment Corporation* (DEC) *Packet Data Protocol* (PDP) minicomputer architecture used by academic and research institutions. In subsequent developments, ARPANET packet technology evolved into the X.25 packet-switching product suite, enabling Internet-designed topologies of commercial and government networks in the 1980s through the 1990s.

What Then Is the Internet?

The easiest way to describe this Internet is that it consists of many interconnected networks, the pieces owned and operated by different operators, linked together. The Internet is not a single entity, but rather is a group of networks made to look like a single network. Reality states more specifically that the Internet is a collection of thousands of networks with no central management or policymaking body. It is linked together by a common set of protocols (TCP/IP) that make it possible for users on any one of the networks to communicate with or use the services located on any of the other networks. The *Internet Engineering Task Force* (IETF) develops technical specifications for the Internet. The specifications are called *Requests for Comment* (RFC). If someone sees a better way of conducting business or access methods on the Internet, the process enables the RFC to be submitted to the IETF for evaluation and implementation.

The current Internet is evolving as a commercial and electronic commerce network for just about every business and residential environment. The commercialization has set the pace for the explosion of connectivity and use. Newer applications have also changed the custodianship and management of the network. The newer players, as shown in Figure 7-13, include

- Local *Internet Service Providers* (ISPs) (local providers, mom-and-pop shops, and so on)
- Regional ISPs (local telephone companies, PREPnet, MIDnet, and others)
- National ISPs (AT&T WorldNet, Sprint, UUNet, Earthlink, and so on)

Figure 7-13
The new Internet.

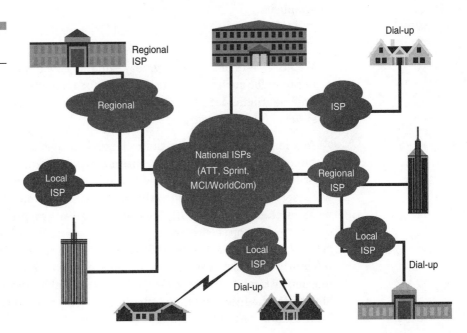

With this commercialization, a wide variety of options became available to gain access to the network. These newer providers have taken over the management responsibility from the government-funded organizations and service providers. In 1995, the *National Science Foundation* (NSF) retired its network and turned it over to these new providers.

Intranets

Larger enterprises have been so impressed with the success and ease of external Internet technology—such as Web servers, Web browsers, and hypertext-based applications—they have now used the same technology internally as their own business communications tool. These organizations have discovered the same features that make the Internet technology attractive for interenterprise communications, also make it an excellent technology for intraenterprise communications. In particular, very large organizations with various applications, hardware platforms, and multiple sources of information can benefit from this technology due to its platform independence. This movement toward intranet technology is supported by other industry trends. For example,

- TCP/IP has become the de facto standard as the multivendor protocol of choice in many corporate networks due to its robustness and simplicity.
- *Simple Mail Transfer Protocol* (SMTP) is rapidly becoming the popular standard for internal e-mail systems.
- Web browsers are becoming very popular as a simple user interface not only for surfing the Net, but also for navigating and accessing organizational data housed on several computer systems and *operating systems* (OSs) software.

At the business-to-business level, the Internet is quickly becoming the most popular means of handling interenterprise communications. Although most businesses continue to use a wide variety of private and public network services, the Internet is the common denominator typically used by most of them. As a result, service and software providers are moving quickly to develop secure, reliable products to enable organizations to transact business over the Internet. Business trust in the Internet has grown significantly over the last year, and some businesses are now transmitting mission-critical data over low-cost, widely available Internet links. Over the long term, the Internet will be the best transport for an organization's communications with its partners.

An internal network is comprised of one or more LANs interconnected with one or more routers running the IP protocol. (So far, this is only an Internet with a lowercase *i*.) The bulk of our traffic on this Net is Web based. That is, we have one or more Web servers and all of the clients are running a browser. Our primary means of communicating is using the Web. (Technically, mail is a different set of application protocols, but most browsers also provide a mail capability.) Perhaps we should say that the primary interface that our users have to their data is via their browser. This network may be geographically far flung where dedicated lines (or Frame Relay) are used to connect our routers. No connection exists to the Internet. This is where things get touchy. If we were to use *Virtual Private Network* (VPN) technology between our corporate locations, one could argue it is still an intranet.

Extranet

By using an external server sitting outside the firewall, the organization has an extranet. Business partners can access the extranet server, and pass or retrieve data from the server. As a query or data request is generated, the server can then spawn a request to the intranet and retrieve that data as needed. By doing so, the server is protected on the inside, but accessible by a limited few from the outside.

The World Wide Web

The proliferation of much of the Internet frenzy occurred when the old UNIX command-line interface was replaced with a *graphical user interface* (GUI) called a browser. As soon as the Internet became friendlier to the non-UNIX user, the demand for and use of data began to explode. Web servers were already deployed throughout the Internet. A plethora of information existed long before the popularity caught up with the technology. However, when the interfaces were developed to enable a user to point and click, the demand to access more systems and more information exploded.

Web-based servers are merely a logical extension of the distributed computing environment, originally the purpose of the Internet. Groups of file servers and database machines exist on the Net. When a user attempts to retrieve information from an organization's home page or Web page, as shown in Figure 7-14, the Web spawns remote procedure calls for the data

Figure 7-14
The World Wide
Web.

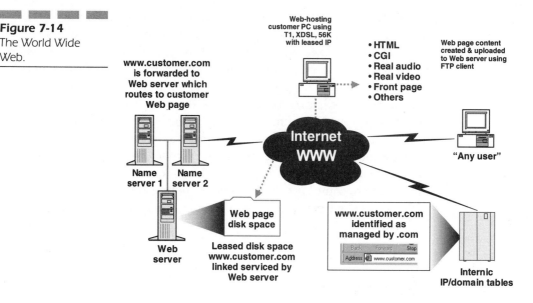

that resides on different machines. As a page is accessed from a local server, the pieces comprising the actual file may be kept elsewhere. Therefore, links and calls to these other servers are built into the Web page using a *Hypertext Transfer Protocol* (HTTP) and a *Hypertext Markup Language* (HTML) to facilitate the ease of use. A call may go to a file server for a small text file, whereas a second request links to a graphic file on a second server. The user sees the final result, without knowing that the data has come from several different places. All the user sees is the completed screen with the appropriate results intended by the owner of the Web page. This, of course, is the reason the screen may fill in at the receiving end in very erratic patterns. As IP delivers the datagrams, TCP determines that it has the file and delivers the information to the terminal or PC. Therefore, as the data arrives from different locations on different routes, the information may appear to be somewhat disjoined. In reality, it is not.

Transmission Control Protocol/Internet Protocol (TCP/IP)

TCP and IP were developed by a U.S. DOD research project to connect a number of networks designed by different vendors. It was successful because it delivered a few basic services that everyone needs (file transfer, electronic mail, and remote logon). Several computers in a small department can use TCP/IP on a single LAN. The IP component provides routing from the department to the enterprise network, then to regional networks, and finally to the global Internet.

As with all other communications protocols, TCP/IP is composed of layers, which are shown in Figure 7-15. This representation is significantly different because it shows the various other protocols that can be contrasted in wired and wireless networks:

- **IP**   Responsible for moving datagrams (commonly referred to as packets) from node to node. IP forwards each datagram based on a four-byte address. The GPRS networks were primarily set up to handle IP routing and switching from a mobile station at higher speeds than were previously available. The Internet authorities assign a range of numbers to different organizations. These organizations assign groups of their numbers to departments. IP operates on gateway machines (actually these are routers) that move data anywhere in the world.

- **TCP**   Responsible for verifying the correct delivery of data from client to server. Data can be lost in the intermediate network. TCP detects errors or lost data and triggers retransmission requests until the data is correctly and completely received.

- **Sockets**   The name given to a package of subroutines that provide access to TCP/IP on most systems.

Figure 7-15
The many protocols in the TCP/IP comparison.

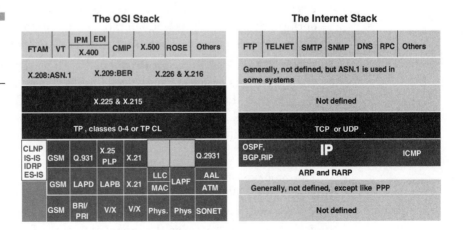

OSI and Internet Protocol Stacks

Internet Protocols (IPs)

The IP was developed to provide internetworking. Individual machines are first connected to a LAN. TCP/IP shares the LAN with other uses (a Novell file server, Windows 95, or Windows NT). A router provides the TCP/IP connection between the LAN and the rest of the world. To ensure that all types of systems from all vendors can communicate, TCP/IP is absolutely standardized on the LAN. Larger networks based on long distances and phone lines are more volatile. Many organizations want to reuse large internal networks based on IBM's *Systems Network Architecture* (SNA). In Europe, the national phone companies traditionally standardized on X.25. However, the explosion of high-speed microprocessors, fiber optics, and digital phone systems has created a burst of new options: *Integrated Services Data Network* (ISDN), Frame Relay, and *Asynchronous Transfer Mode* (ATM). New technologies come and go over a few years. With cable TV and phone companies competing to build the national Information Superhighway, no single standard can govern nationwide or worldwide communications.

The original design of TCP/IP fits nicely within the current technological uncertainty. TCP/IP data can be sent across a LAN, can be carried within an internal corporate SNA network, or can piggyback on the cable TV. Furthermore, machines connected to any of these networks can communicate to any other network through gateways supplied by the network vendor. Several other protocols are used with TCP/IP. These are usually bundled together into the TCP/IP protocol suite. These other protocols include the following:

- **User Datagram Protocol (UDP)**   The UDP is used with applications that do not need to sequence datagrams. UDP does not perform as many checks and balances on the data as TCP does. It does not keep track of what is sent segment the data like TCP.

- **Internet Control Message Protocol (ICMP)**   The ICMP is used for error messages and other messages intended for TCP/IP software itself, rather than any particular user program. ICMP is similar to UDP. It fits all its information into a single datagram and therefore does not have to break it down into many datagrams. ICMP works at the same layer as IP.

The major reason for TCP/IP's success is that it can be ported across multiple platforms. Whereas other protocol stacks are typically used in a LAN or a WAN environment, TCP/IP works on all aspects of the internetwork. A typical network may involve the integration of the following:

- The LAN operating on a Token Ring or an Ethernet. The LAN OS may be Novell, Windows 95, Windows NT, or other.

- The Campus Area Network using *Fiber-Distributed Data Interface* (FDDI) within a campus or high-rise office building.

- The Metropolitan Area Network linking networks with either a leased line (such as T1, T3, or SONET) or *Switched Multimegabit Data Services* (SMDS).

- The WAN using Frame Relay, ISDN, or ATM and now GPRS and *High-Speed Circuit-Switched Data* (HSCSD).

Regardless of the topologies, TCP/IP is robust enough to link all the various systems together without any proprietary protocols. IP uses several different ways of handling the data transmission across a network. Unlike the OSI model, the X.25, the circuit-switched or leased-line networks where a connection is established between the two endpoints, IP uses a connectionless-oriented protocol. As a connectionless-oriented service, IP does not establish a true connection with the far end receiving the data.

IP sends datagrams into the network with a best attempt to deliver the receiving end. IP does not know if the device exists or if it is online. IP makes no guarantees that the data will ever be delivered. IP does not concern itself with the integrity of the data. In X.25, every node processing packets checked the data integrity and the sequencing. IP does not. It will deliver bad or corrupted data to the far end, if the datagram is deliverable at all.

IP does manage to route data through and to dissimilar networks. Whether the user is running a mainframe with SNA or a UNIX platform, IP does not get involved with the data differentiation. Another difference between IP and other protocols is that IP does not deal with the correct sequencing of the datagrams. Every datagram in the IP world is its own entity. Datagrams are not numbered 1, 2, 3, and so on, as one will find in the X.25 packet-handling mechanisms and protocols. For all intents and purposes, IP is referred to as a dumb protocol because it does nothing except attempt to deliver the data. TCP sends its segments to IP. All it really tells IP about the data is the address. IP does not consider content. Its job is simply to get the information to the far end. IP also is told what protocol is used for delivery. Because other protocols can use IP, even though the bulk of the data uses TCP, it is necessary to differentiate the protocol being used inside the datagrams. IP does add some overhead to the datagrams as it passes the data onto the network.

TCP

TCP is responsible for breaking the actual data up into segments. Figure 7-16 shows the TCP header. The end-user application may be sending large files from one host to another. TCP breaks the larger files into more manageable pieces called segments. TCP also reassembles the segments back into the original message at the receiving end, arranged in its proper order. It may seem like TCP is doing all the work on the TCP/IP network. This is mostly true in smaller networks. However, when dealing with larger networks and the Internet, routing the data from one end to another is no trivial task. TCP identifies which connection the segment is part of by using a connection-oriented protocol. This task of keeping track of the incoming segments and delivering them to the right connection (application) is called the demultiplexing step. TCP uses its own header information to keep track of the connection. The header is also used to define how large the segments will be as the two end systems pass information back and forth. TCP has to know how much the far end can handle. The TCP protocol at each end describes the size of the segment, and then they mutually agree to use the smallest size for both systems. Typically, the datagrams in IP are 576 octets, although this is not a given requirement. The datagrams can actually be as large as 64KB large, but few organizations use that size. The maximum transmission unit defines how large the datagrams will be and how many datagrams will be used in a segment.

Figure 7-16
The TCP header.

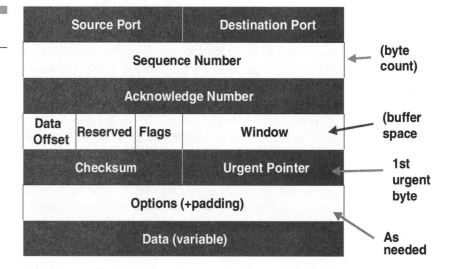

TCP ensures the integrity of the data when it arrives, as well as reordering the data back into sequence. If data is corrupted, TCP asks for a retransmission. If data does not arrive, or if it does not arrive within a specified period of time (usually one second), TCP requests a retransmission. TCP does not number its segments. Instead, it numbers the octets to send. It performs the ACK/NAK function based on octet counts expected versus received.

Address Resolution Protocols (ARPs)

On some media (such as IEEE 802 LANs), media addresses and IP addresses are dynamically discovered through the use of two other members of the IP suite:

- *Address Resolution Protocol* (ARP)
- *Reverse Address Resolution Protocol* (RARP)

ARP uses broadcast messages to determine the hardware *Media Access Control* (MAC)-layer address corresponding to a particular internetwork address. ARP is sufficiently generic to enable the use of IP with virtually any type of underlying media-access mechanism. When the MAC address is not known, ARP generates a message across the LAN, as shown in Figure 7-17. The message is that the sender has an IP address to send, but it does not know the MAC address. This broadcast asks for the MAC layer address to use. All devices on the network hear the broadcast, but only the

Figure 7-17
The ARP query.

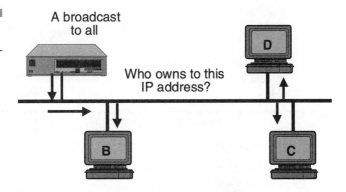

Address Resolution Protocol (ARP) 1

device with the IP address responds. It sends back a message to use the following MAC address, as shown in Figure 7-18. All other devices on the subnet hear the request and the response, and then update a table in their appropriate memory.

RARP uses broadcast messages to determine the Internet address associated with a particular hardware address, as shown in Figure 7-19. RARP is particularly important to diskless nodes, which may not know their internetwork address when they boot. The response is sent, which is shown in Figure 7-20.

Figure 7-18
The ARP response.

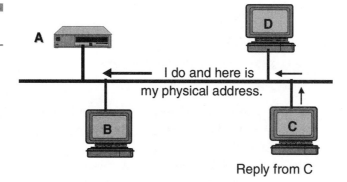

Address Resolution Protocol (ARP)-2

Figure 7-19
The RARP query.

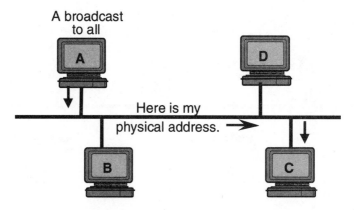

RARP Operations   Query- 1

Figure 7-20
The RARP response.

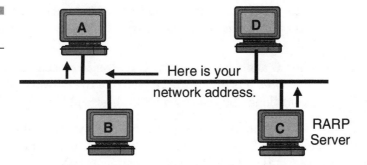

RARP Operations Response-2

IP Addressing

Each address assigned to a network is given an IP address consisting of a four-byte address. The addressing mechanisms are controlled by an organization called the *Internet Network Information Center* (InterNIC). Every Internet address is comprised of a network and a host address. The InterNIC assigns a class of network address that is broken down into the following classes, as shown in Figure 7-21:

■ Class A addresses use the first 8 bits (the first bit is set to a 0, leaving 7 bits for the network number) of the address for the network number, followed by a 24-bit address for the host.

Figure 7-21
IP addressing.

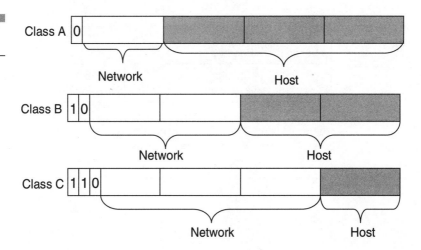

■ Class B addresses use a 16-bit network number (the first two bits are reserved and set to 10), leaving a 14-bit network number and a 16-bit host address.

■ Class C addresses use a 24-bit address for the network number (the first three bits are set to 110, leaving the network address field at 21 bits long) followed by an 8-bit host address.

Like anything else in a network so widely deployed worldwide, a problem looms in the future. We are running out of address numbers! In the newest version of IP, called IP New Generation or IP Version 6, the address field is expanded to a 128-bit address field for every source and destination node on the network. Figure 7-22 shows the IP Version 6 header.

IP Subnetworking and Masking

When using TCP/IP on a network, a subnetwork address and a masking of the network and host addresses can be used. The subnetwork address can be assigned on a departmental basis. Using subnetworking, an organization can use a class B address and then subnetwork each individual LAN or department with a class C address, as shown in Figure 7-23. The subnet address can advise the routing devices to ignore the parts of the network not needed to route the data from one network to another. This is more efficient than having to use the entire address when searching the routing tables. If a network administrator has chosen to use 8 bits of subnetting, the third octet of a class B IP address provides the subnet number. For exam-

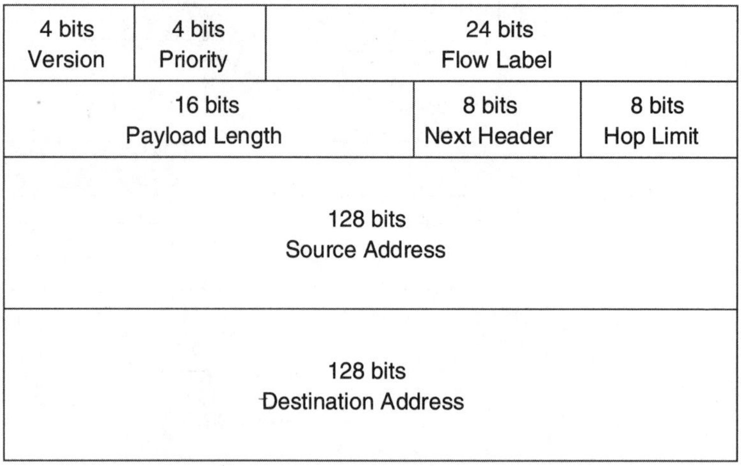

Figure 7-22
IPv6 headers.

4 bits Version	4 bits Priority	24 bits Flow Label	
16 bits Payload Length		8 bits Next Header	8 bits Hop Limit
128 bits Source Address			
128 bits Destination Address			

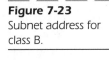

Figure 7-23
Subnet address for class B.

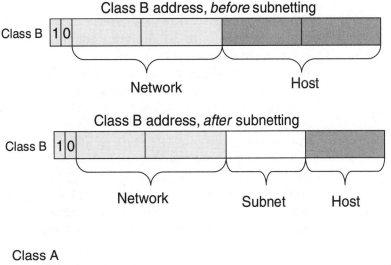

Class B address, *before* subnetting

Class B address, *after* subnetting

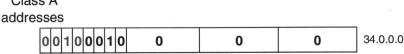

Figure 7-24
A subnet mask for class A address.

ple, address 192.168.1.0 refers to network 192.168, subnet 1; address 192.168.2.0 refers to network 192.168, subnet 2; and so on.

The number of bits borrowed for the subnet address varies. To specify how many bits are used, IP provides the subnet mask. Subnet masks use the same format and representation technique as IP addresses. Subnet masks have ones in all bits except those bits that specify the host field. For example, the subnet mask that specifies 8 bits of subnetting for class A address 34.0.0.0 is 255.255.0.0, as shown in Figure 7-24.

Internet Routing

Routing devices in the Internet have traditionally been called *gateways*—an unfortunate term because elsewhere in the industry the term applies to a device with somewhat different functionality. Gateways (which we will call routers from this point on) within the Internet are organized hierarchically. Some routers are used to move information through one particular group of networks under the same administrative authority and control (such an entity is called an autonomous system). Routers used for information exchange within autonomous systems are called interior routers, and they use a variety of *Interior Gateway Protocols* (IGPs) to accomplish this purpose. Routers that move information between autonomous systems are called exterior routers, and they use *Exterior Gateway Protocols* (EGPs) for this purpose.

IP routing protocols are dynamic. Dynamic routing calls for routes to be calculated at regular intervals by software in the routing devices. This contrasts with static routing, where routes are established by the network administrator and do not change until the network administrator changes them. An IP routing table consists of destination address/next-hop pairs. A sample entry is interpreted as meaning "to get to network 34.1.0.0 (subnet 1 on network 34), the next stop is the node at address 54.34.23.12." Figure 7-25 shows a sample routing table.

IP routing specifies that IP datagrams travel through an internetwork one hop at a time. The entire route is not known at the outset of the journey. Instead, at each stop, the next destination is calculated by matching the destination address within the datagram with an entry in the current node's routing table. Each node's involvement in the routing process consists only of forwarding datagrams based on internal information, regard-

Figure 7-25
A routing table.

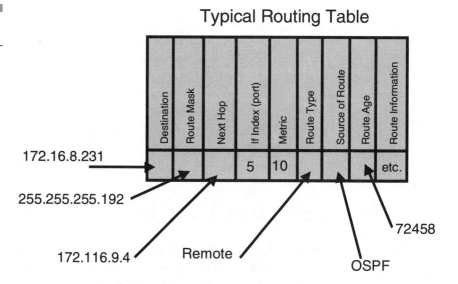

Typical Routing Table

Destination	Route Mask	Next Hop	If Index (port)	Metric	Route Type	Source of Route	Route Age	Route Information
172.16.8.231	255.255.255.192	172.116.9.4	5	10	Remote	OSPF	72458	etc.

less of whether the datagrams get to their final destination. In other words, IP does not provide for error reporting back to the source when routing anomalies occur. This task is left to another IP, the ICMP.

ICMP

ICMP performs a number of tasks within an IP internetwork. The principle reason it was created was for reporting routing failures back to the source. In addition, ICMP provides helpful messages such as the following:

- Echo and reply messages to test node reachability across an internetwork
- Redirect messages to stimulate more efficient routing
- Time exceeded messages to inform sources that a datagram has exceeded its allocated time to exist within the internetwork
- Router advertisement and router solicitation messages to determine the addresses of routers on directly attached subnetworks

A more recent addition to ICMP provides a way for new nodes to discover the subnet mask currently used in an internetwork. All in all, ICMP is an integral part of any IP implementation, particularly those that run in routers.

IRDP

The *ICMP Router Discovery Protocol* (IRDP) uses router advertisement and router solicitation messages to discover addresses of routers on directly attached subnets. In IRDP, each router periodically multicasts router advertisement messages from each of its interfaces. Hosts discover the addresses of routers on the directly attached subnet by listening for these messages. Hosts can use router solicitation messages to request immediate advertisements, rather than waiting for unsolicited messages. IRDP offers several advantages over other methods of discovering addresses of neighboring routers. Primarily, it does not require hosts to recognize routing protocols, nor does not it require manual configuration by an administrator. Router advertisement messages enable hosts to discover the existence of neighboring routers, but not which router is best to reach a particular destination. If a host uses a poor first-hop router to reach a particular destination, it receives a redirect message identifying a better choice.

Transport Layer

TCP and the UDP implement the Internet transport layer. TCP provides connection-oriented data transport, whereas UDP operation is connectionless.

Transmission Control Protocol (TCP)

TCP provides full-duplex, acknowledged, and flow-controlled service to upper-layer protocols. It moves data in a continuous, unstructured byte stream where sequence numbers identify bytes. TCP can also support numerous simultaneous upper-layer conversations.

TCP Segment Format

The fields of the TCP segment are as follows:

- **Source port and destination port**   Identifies the points at which upper-layer source and destination processes receive TCP services.
- **Sequence number**   Usually specifies the number assigned to the first byte of data in the current message. Under certain circumstances, it can also be used to identify an initial sequence number to be used in the upcoming transmission.
- **Acknowledgment number**   Contains the sequence number of the next byte of data the sender of the packet expects to receive.
- **Data offset**   Indicates the number of 32-bit words in the TCP header.
- **Reserved**   Reserved for future use.
- **Flags**   Carries a variety of control information.
- **Window**   Specifies the size of the sender's receive window (that is, the buffer space available for incoming data).
- **Checksum**   Indicates whether the header was damaged in transit.
- **Urgent pointer**   Points to the first urgent data byte in the packet.

- **Options**   Specifies various TCP options.
- **Data**   Contains upper-layer information.

User Datagram Protocol (UDP)

UDP is a much simpler protocol than TCP and is useful in situations where the reliability mechanisms of TCP are not necessary. The UDP header has only four fields: source port, destination port, length, and UDP checksum. The source and destination port fields serve the same functions as they do in the TCP header. The length field specifies the length of the UDP header and data, and the checksum field provides packet integrity checking. The UDP checksum is optional. Figure 7-26 shows the UDP header.

Upper-Layer Protocols

The IP suite includes many upper-layer protocols representing a wide variety of applications, including network management, file transfer, distributed file services, terminal emulation, and electronic mail. The best-known Internet upper-layer protocols support certain applications.

The *File Transfer Protocol* (FTP) provides a way to move files between computer systems. Telnet enables virtual terminal emulation. The *Simple Network Management Protocol* (SNMP) is a network management protocol used for reporting anomalous network conditions and setting network threshold values. X Windows is a popular protocol that permits intelligent terminals to communicate with remote computers as if they were directly attached. *Network File System* (NFS), *External Data Representation* (XDR), and *Remote Procedure Call* (RPC) combine to provide transparent access to

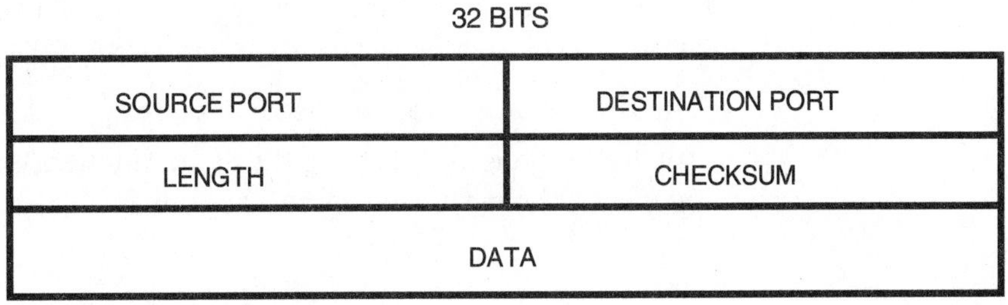

Figure 7-26
UDP header.

```
                              32 BITS

| SOURCE PORT        | DESTINATION PORT   |
| LENGTH             | CHECKSUM           |
|            DATA                         |
```

remote network resources. The SMTP provides an e-mail transport mechanism. These and other network applications use the services of TCP/IP and other lower-layer IPs to provide users with basic network services.

■ Network file services give access to files and data on remote or disparate network servers.

■ Remote printing capability enables a user to send print files to a remote printer as though it were a local attached printer.

■ Remote program execution enables a user to access and launch a program on a remote computing system, appearing as a locally attached terminal device.

■ Name services uses the ARPs and RARPs to determine the addressing on a particular device on a network and host mechanism. Name services also include the translation from a unique naming convention to an IP address (for example, bud@tcic.com translates to 198.120.205.10).

■ Networking windows systems enable the TCP stack to link devices as peers on a network accessible using a GUI.

■ Terminal services enable the user to emulate a dumb terminal and access data in a form and format acceptable by the remote host.

■ E-mail access for a SMTP makes e-mail a simple procedure between different systems.

The IP Header

The IP Version 4 header is attached to the datagram to get the information to the destination address, as shown in Figure 7-27. The advantages of the robustness of the TCP/IP world are based on IP's capability to deliver the data across a wide area of disparate systems and platforms. By using the header, containing the addressing and other pieces of critical information, a trailer is not needed to move the information. The header consists of the following pieces of information (not all are covered):

■ The IP version number of the IP datagram.

■ The header length specifying the datagram header length in units of 32 bit words; the most common is 20 octets.

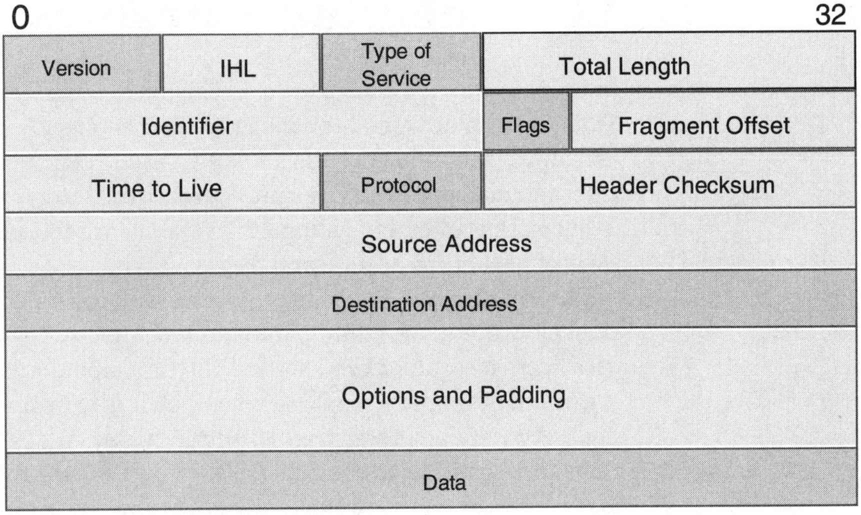

Figure 7-27
IPv4 header.

- The type of service covers the quality of service and is used when different algorithms are used for routing the datagrams across the network. Unfortunately, the entire industry uses this field differently so a standard is not defined.
- The total length defines the length of the entire datagram and overhead.
- The identification, flags, and offsets control the fragmentation of the datagrams and is used for the reassembly portion of the process.
- The *time-to-live* (TTL) is a hop counter that uses a decrementing counter for the delivery of the datagram. How many devices may the datagram pass through before being either delivered or discarded?
- The Protocol field is used to define the higher-level protocol used (TCP or UDP).
- The Header checksum is a CRC on the header for error detection of any problems with the datagram.
- Next, the source and destination addresses show.
- Options include any special features allowed on the network.
- Padding is filler used to align the datagram and header to a 32-bit alignment.

Implementing Extranets

In order to provide some sort of consistency, some definitions are required. These are not universal (and are violated frequently in the trade press) but help with the following discussions. Organizations are connected to the Internet and specifically permit access to the internal network (or portions of our network) by its customers. Now the organization is at risk from all sorts of hackers, crackers, and freaks.

The goal then is to keep the outsiders (hackers and crackers) out, or at least slow them down and make them work for it, while giving friends and customers free, unfettered access. The first step then is to install a firewall. A firewall is simply a machine whose job it is to examine each packet arriving and verify whether it is permitted based on a set of rules. This set of rules is referred to as the rule base. Firewalls essentially operate at the IP and TCP levels of the protocol stack. This means that the firewall examines each packet to see if it is coming from and going to the proper IP address and that it contains the proper or permitted TCP port number. These firewalls can be set up as a one-way valve. Employees can surf the Net at will and send out any kind of packet, but the firewall will not permit any unsolicited packets. (This means that mail won't work; that is, the firewall won't let in any packets including mail.) The other problem with this approach is that it will permit internal employees to hack on the Net. If they are caught, the corporation may be liable so think about this issue. This is not to say that firewalls are not useful but to point out that simplistic solutions will not achieve the organization's objectives.

Performing basic packet filtering on a source and destination IP address is a useful feature when combined with additional firewall capabilities. It was initially believed that a firewall was the solution to all our Internet security problems. The limitations of a firewall's only implementation are now apparent. If we only need to communicate between specific networks or specific devices with fixed IP addresses, a firewall is a fine solution.

TCP Filtering

The next level of checking that the firewall provides is to check inside the TCP packet for port number. For example, the standard TCP port for HTTP is port 80. (Any port can be could be used in this example. Port 8080 is another port frequently used by proxies.) The firewall now checks for IP addresses (which might say any IP address) and the port number in the TCP packet.

This means that unless the port is open through the firewall, service cannot be reached (such as HTTP or FTP). This provides for a more general audience. Now instead of filtering on an IP address one can say let anybody from any IP address come through the firewall, but only if he or she is doing HTTP on port 80 to a specific server. This means that other Web pages may be on other servers on port 80 internally, but they would not be available because the firewall limits access to a specific server. Clearly, access may be granted to as many servers as desired.

This presentation is a little simplistic because it makes the assumption that the OS on which the firewall software is running is itself secure and unhackable. Not only is this not generally true, but it is impossible to secure the OS perfectly. (For example, attacks have occurred to a UNIX-based system that is running send mail. Therefore, the send mail application should not be present on your firewall machine. One only has to read the trades to see the attacks and open doors being discovered in all the OSs.) Thus, it isn't just enough to run a firewall; the firewall must be run on a locked-down OS.

Two general kinds of attacks against systems exist. One is called the denial of service attack. In this case, the hacker sends one or thousands of packets to the server, causing it to spend all its processing power processing these garbage packets so that it has no time left over to do its real job, such as serving up Web pages. The second kind of attack is more invasive. In this case, the attacker tries to gain access to the machine preferably as the superuser. If the attacker achieves this, he or she can do anything that he or she wants. As a superuser, the attacker can set up accounts on the machine for his or herself and his or her friends, read any file, wipe out any file, and so on. Very clever hackers will therefore edit the log files so that these logs show no trace of the hacker's presence. You can keep log file in a couple of ways. One way is to frequently copy the log file to an obscure directory and change its name so that its name does not contain "log." The second is to encrypt the file. The latter is only convenient if the firewall application provides this option. The former only works if the log file is copied often enough to capture the hacker's presence before he or she cleans the log file.

In general, care must be taken with the use of a packet network and more specifically, the IP protocol on the Internet. In a wireless environment, this is equally critical because the end user is somewhat transparent. The *Gateway GPRS Support Node* (GGSN) in a GPRS network makes the mobile station's mobility transparent. This just opens the door to more attacks and more complicated solutions.

The combination of the packet switching technologies (X.25 and IP) make the transfer of data communications a smooth operation in today's networks. Although X.25 has waned in its popularity, it is still used throughout the world in emerging countries that have older networks in place. The use of IP makes use of the connectionless transfer of data more robust and simplifies its implementation. Nearly every computer manufacturer today has embedded IP in the OS as a core protocol element.

GPRS takes advantage of the standard interfaces that have evolved over the years to support the packet data networks through these simple interfaces. That way, GPRS can be implemented as an on-ramp service to a public data network, rather than having to build a whole new architecture or set of protocols. Having spent the time to consider the use of public X.25 networks and the Internet, coupled with the intranet and extranet, we now have a basis on which the applications can be supported from a hand-held wireless interface. In Chapter 11, "Future Enhancements and Services," these applications of the intranet and Internet integartion culminate in some of the application-specific uses for GPRS in the future.

Mobile Station to SGSN Interface

Objectives

Upon completion of this chapter, you should be able to

- Understand the way the mobile station communicates with the SGSN via LLC.
- Discuss the ways the data protocols work at Layer 2 and 3.
- Describe the use of SNDCP protocols.
- Understand the services that the SGSN offers to the mobile station.
- Describe how the protocols stack up.
- Describe the use of the TLLI.

Logical Link Control (LLC) Layer

The *Logical Link Control* (LLC) layer provides a reliable logical link between the *mobile station* (MS) and the *Serving GPRS Support Node* (SGSN). A *Temporary Logical Link Identifier* (TLLI) is used for addressing at the LLC layer. The LLC is independent of the underlying radio interface protocols. The LLC provides services necessary to maintain a ciphered data link between a mobile station and an SGSN.

The logical link is maintained as the mobile station moves between cells served by the same SGSN. When the mobile station moves to a cell being served by a different SGSN and performs a new attach, the existing connection is released and a new logical link connection is established (this is done by the attach operation and the *Packet Datagram Protocol* [PDP] context activation). The LLC provides for acknowledged and unacknowledged point-to-point delivery of LLC *protocol data units* (PDUs) between the mobile station and the SGSN and point-to-multipoint delivery of packets from the SGSN to the mobile station.

The LLC layer also provides procedures for detecting errors from corrupted PDUs. It does this by checking the *frame check sequence* (FCS) in the LLC frame format. The FCS contains the value of the *Cyclic Redundancy Check* (CRC) calculation that is performed over the entire contents of the header and the information fields. The unacknowledged mode of transfer has no error recovery. For the acknowledged mode of transfer, the LLC may request retransmission of the frames of data for which an acknowledgment has not been received.

What LLC Does

This layer provides a highly reliable ciphered logical link. A logical link between a mobile station and its SGSN is identified by a TLLI. The LLC is independent of the underlying radio interface protocols in order to accept the introduction of alternative GPRS radio solutions with minimum changes to the *Network Subsystems* (NSS). These added radio solutions could include techniques such as *Code Division Multiple Access* (CDMA) or ANSI-136+ from the North American standards. The movement to a *Universal Mobile Telecommunications System* (UMTS) is another possibility where the *Universal Terrestrial Radio Access Network* (UTRAN) is introduced. The LLC connection can be used to transfer point-to-point and point-to-multipoint data between the mobile station and the SGSN. LLC is designed to support many Layer 3 protocols, such as those shown in Figure 8-1.

Subnetwork-Dependent Convergence Protocol (SNDCP)

Network layer protocols are intended to operate over services derived from a wide variety of subnetworks and data links. GPRS supports several network layer protocols providing protocol transparency for users of the service. The introduction of new network layer protocols to be transferred over GPRS is possible without any changes to GPRS. This is the role of the *Subnetwork-Dependent Convergence Protocol* (SNDCP) layer.

Figure 8-1
The LLC supports multiple protocols at Layer 3.

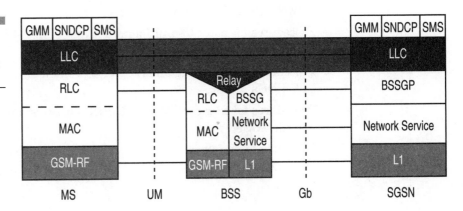

GPRS Mobility Management/Session Management (GMM/SM)

GPRS Mobility Management / Session Management (GMM/SM) protocol supports *mobility management* (MM) functionality such as attach and authentication, and transport of session management messages for functions such as PDP context activation and deactivation.

Short Message Service (SMS)

The *Short Message Service* (SMS) uses the services of the LLC layer to transfer short messages between the mobile station and the SGSN.

LLC Support

The LLC layer supports the following functions:

- Service primitives enabling the transfer of *SNDCP protocol data units* (SN-PDUs) between the SNDCP layer and the LLC layer
- Procedures for transferring *LLC protocol data units* (LLC-PDUs) between the mobile station and the SGSN
- Procedures for unacknowledged point-to-point delivery of LLC-PDUs between the mobile station and the SGSN
- Procedures for acknowledged point-to-point delivery of LLC-PDUs between the mobile station and the SGSN
- Procedures for point-to-multipoint delivery of LLC-PDUs from the SGSN to the mobile station
- Procedures for detecting and recovering from lost or corrupted LLC-PDUs
- Procedures for flow control of LLC-PDUs between the mobile station and the SGSN
- Procedures for ciphering the LLC-PDUs (applies to both unacknowledged and acknowledged LLC-PDU delivery)

The layer functions are organized so that the ciphering resides immediately above the *Radio Link Control / Medium Access Control* (RLC/MAC) layer in the mobile station and immediately above the *BSS GPRS Protocol* (BSSGP) layer in the SGSN.

LLC Service Access Point Identifiers (SAPIs)

The LLC provides six *Service Access Points* (SAPs) to the upper-layer protocols. Think of these as tunnels between the layers. The tunnels enable the passage of data between the Layer 2 and Layer 3 entities, as shown in Figure 8-2 and described as follows:

- Four SAPs are dedicated to the SNDCP that manages data packet transmission. This protocol only deals with transmission. Currently, only one SAP exists for each *quality of service* (QoS).

- One SAP is dedicated to GMM.

- One SAP is dedicated to the SMS.

A *Service Access Point Identifier* (SAPI) identifies each SAP.

Figure 8-2
The LLC SAPI functions as an access to the Layer 3 protocols.

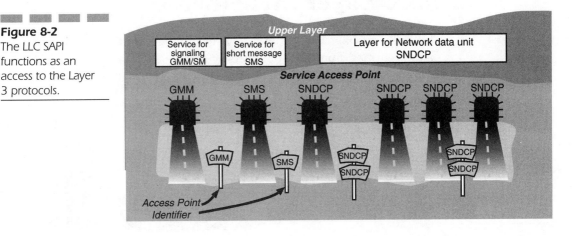

LLC Identifiers

A logical link is unique to a particular mobile station. A *Data Link Connection Identifier* (DLCI) identifies this logical link. A DLCI is composed of the SAPI at the LLC and the TLLI. Figure 8-3 shows the LLC identifier.

SAPIs are points at which the LLC provides access to the SNDCP layer (that is, the *Network Service Access Point Identifiers* [NSAPIs]). The SAPIs that the LLC provides to the SNDCP layer are essentially the four QoS levels provided for data communication for different levels of reliability. More than one NSAPI may be using a particular SAPI depending upon the type of reliability required by that service. Several NSAPIs may request the same type of QoS level and hence, an SAPI at the LLC can support multiple NSAPIs from the SNDCP layer.

The NSAPIs access the SAPIs and the SAPIs are grouped together with the TLLI that is specific for a particular mobile station. In wireless packet data communication, the physical link does not need to be maintained for the entire duration of the call. The mobile may be in the idle state. It transitions back to the active state when it has to send a packet or when it is going to receive packets. The logical link connection is maintained even when the lower layers no longer exist. This means that radio resources are not used until needed, keeping the goal of the always-there and always-on data transmission without consuming the operators' resources.

LLC Layer Structure

In the functional model of the LLC layer, different functions are given to different entities. Functions provided by each *Logical Link Entity* (LLE) are shown in Figure 8-4 and are as follows:

- Unacknowledged and acknowledged information transfer

Figure 8-3
The LLC identifiers
create a DLCI.

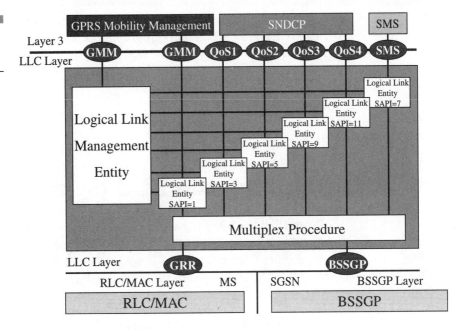

Figure 8-4
The functions of the
LLEs.

- Flow control
- Frame error detection

The LLE analyzes the control field of the received frame and provides appropriate responses and layer-to-layer indications. Functions of the multiplex procedure in the LLC layer are as follows:

- On frame transmission,
1. Generates and inserts the FCS
2. Performs the frame ciphering function
3. Provides SAPIs

- On frame reception,
1. Performs the frame decipher function and checks the FCS
2. Distributes the frame to the appropriate LLE
3. Performs GPRS deciphering function

The *Logical Link Management Entity* (LLME) manages the resources that have an impact on individual connections. One LLME exists per TLLI. Functions provided by the LLME are

- Initializing the parameters to be used
- Error processing
- Invoking connection flow control

Mapping the LLC Frame

You can find the SAPI parameter in the address field of the LLC frame. Figure 8-5 shows the different values the SAPI can take in function of the service required. Note that TLLI is not transmitted in LLC frames. The length of the LLC header depends on the frame format:

- The address field is fixed at 1 octet.
- The FCS field is fixed at 3 octets.
- The control field depends on transfer mode being used as follows:

    In unacknowledged mode,

    - The Information frame (format UI) has a 2-octet control field length.
    - The Control frame (format U) has a 1-octet control field length.

Figure 8-5
The information mapping on the LLC frame.

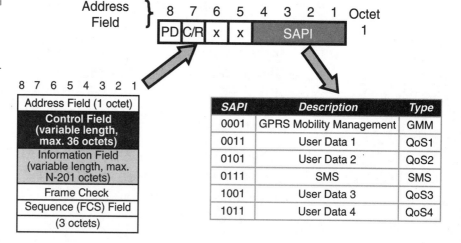

SAPI	Description	Type
0001	GPRS Mobility Management	GMM
0011	User Data 1	QoS1
0101	User Data 2	QoS2
0111	SMS	SMS
1001	User Data 3	QoS3
1011	User Data 4	QoS4

In acknowledged mode,

- The Information frame (format I) has a 3-octet control field length plus a possibility of an acknowledgment bitmap (max of 36 octets)
- The Supervisory frame (format S) has a 2-octet control field length plus a possibility of an acknowledgment bitmap (max of 34 octets).

GPRS Ciphering Environment

In normal operation, the use of the radio link is considered both hostile (noisy and error prone) and a security risk in that it is open to active interception techniques. To prevent the unauthorized reception in the radio link, security in GSM networks is based on two primary techniques:

- Authentication
- Ciphering (encryption)

Authentication

The *Authentication Center* (AuC) is responsible for generating a set of parameters known as triplets. A triplet consists of a

- *Cipher Key* (K_c)
- *Random Number* (RAND)
- *Signed Response* (SRES)

The RAND is a randomly generated number from a number pool containing 2^{128} numbers. The RAND, coupled with the *Identification Key* (K_i), is used to calculate K_c and SRES. K_i is a secret number allocated on a per-subscriber basis and is only held at the AuC and is based on the *Subscriber Identity Module* (SIM) card. Measures are taken to ensure that the K_i cannot be read from the SIM card. K_i is never transmitted over the network.

Ciphering

Ciphering is used over the air interface following the authentication procedure to provide security for voice and data traffic. *Algorithm 5* (A_5) is used with K_c and the current *Time Division Multiple Access* (TDMA) frame number as inputs to generate a ciphering code. The mobile station calculates K_c from the RAND and K_i and stores it on the SIM. The *Base Station Subsystem* (BSS) is given K_c by the *Visitor Location Register* (VLR). In the uplink direction, the mobile ciphers the data and the BSS deciphers it. A similar process takes place on the downlink. The cipher key is different in the uplink and downlink direction. The TDMA frame number changes approximately every 4.6 ms (a TDMA frame period) and is not repeated for 3.5 hours, making it difficult for the cipher code to be cracked. Some countries allow ciphering as an option; others forbid it. Figure 8-6 shows this sequence.

In GPRS, the ciphering is no longer performed between the *Base Transceiver System* (BTS) and the mobile station, but at the LLC level between the SGSN and the mobile station. Just as we saw in GSM, a ciphering key K_c is produced by the A_8 algorithm from the identification key K_i and a nonpredictable RAND is generated by the AuC and transmitted by the network to the mobile station.

Each time a mobile station is authenticated, the A_5 algorithm is used to calculate the expected sequence used for ciphering and deciphering in the mobile station and in the SGSN. The A_5 algorithm is stored in the mobile equipment and in the SGSN. Thus, the A_5 is manufacturer-dependent. This is shown in Figure 8-7.

Figure 8-6
The sequence in GSM ciphering.

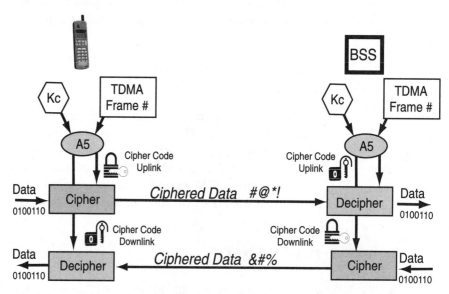

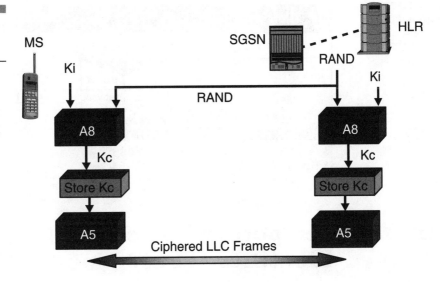

Figure 8-7
The GPRS ciphering environment.

The performance requirements on the GPRS ciphering algorithm are expected to be similar to those of the existing A_5 algorithm from the GSM architecture. The main difference is that ciphering for uplink and downlink transfers are independent. Therefore, the ciphering algorithm may be different.

GPRS Mobility Management (GMM)

GMM uses the services of the LLC layer to transfer messages between the mobile station and the SGSN. GMM includes functions such as GPRS attach, GPRS detach, security, cell update, *routing area* (RA) update, location update, PDP context activation, and PDP context deactivation. In addition, GMM functions support the management of the three states (ready, standby, and idle) on both sides (mobile station and SGSN). The GMM informs the network of the whereabouts of the mobile station and provides confidentiality for the user. These are crucial functions that the mobility and session management functions operate at Layer 3. As Figure 8-8 shows, the GMM layers are transparent to the underlying layers and the BSS.

Figure 8-8
GMM layer.

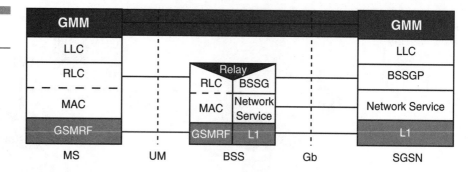

Diagram columns: MS | UM | BSS | Gb | SGSN

Temporary Logical Link Identifier (TLLI)

Within an RA, a one-to-one relationship exists between the TLLI and the *Temporary Mobile Subscriber Identity* (TMSI) (an *International Mobile Subscriber Identity* [IMSI]) that is only known in the mobile station and the SGSN. Whereas a *packet TMSI* (P_TMSI) is used in the GMM sublayer for identification of a mobile station, a TTLLI is used for addressing the radio resources. TLLI is also derived from the P_TMSI and unambiguously identifies the logical link.

■ TLLI assignment is controlled by the GMM.

■ TLLI is assigned, changed, and unassigned with an assignment request function coming from interaction between the LLC (more precisely in the LLME) and the GMM layer.

The mobile station will transmit its IMSI (only when necessary, otherwise the temporary IMSI is used whenever possible for security reasons) or the old P_TMSI to the SGSN when attaching to the SGSN. On attachment, the mobile station uses the P_TMSI for authentication. The SGSN uses the old P_TMSI to reallocate a new one. Therefore, a new P_TMSI will be assigned to the mobile station and passed across the network to the mobile station. The mobile station having received the new P_TMSI from the network transmits a TLLI to unambiguously describe the logical link between it and the network. This is the one-to-one relationship previously described. The P_TMSI is coded with a 32-bit sequence to create the TLLI, as shown in Table 8-1.

Table 8-1

The Format of the TLLI Mapped in 32 Bits from the P_TMSI

Type of TLLI	0–26	27	28	29	30	31
Local TLLI			P_TMSI		1	1
Foreign TLLI			P_TMSI		0	1
Random TLLI	R	1	1	1	1	0
Auxiliary TLLI	A	0	1	1	1	0
Reserved	X	X	0	1	1	0
Reserved	X	X	X	0	1	0
Reserved	X	X	X	X	0	0

How the TLLI Is Used

A TLLI is used for addressing on resources used for GPRS. This TLLI is built on the basis of either the local or foreign P_TMSI. This means that a valid P_TMSI is available. If the mobile station has stored a valid P_TMSI in the SIM, the mobile station derives a foreign TLLI from that P_TMSI and uses it for transmission of the following messages:

- ATTACH REQUEST message of any GPRS attach procedure
- RA UPDATE REQUEST message procedure if the mobile station has entered a new RA

Any other GMM message is transmitted using a local TLLI derived from the P_TMSI or directly using a random TLLI, which means no valid P_TMSI is available. When the mobile station does not have a valid P_TMSI stored (that is, the mobile station is not attached to GPRS), the mobile station may use a randomly selected random TLLI for transmission of the ATTACH REQUEST message of any GPRS attach procedure.

Upon receipt of an ATTACH REQUEST message, the network assigns a P_TMSI to the mobile station, derives a local TLLI from the assigned P_TMSI, and transmits the assigned P_TMSI to the mobile station.

Upon receipt of the assigned P_TMSI, the mobile station derives the local TLLI from this P_TMSI and uses it for addressing at lower layers.

In both cases, the mobile station acknowledges the reception of the assigned P_TMSI to the network. After receipt of the acknowledgement, the network uses the local TLLI for addressing at lower layers.

Yet, another type of TLLI is available called the anonymous access request (auxiliary TLLI). In order to activate an anonymous PDP context, the mobile station sends an ACTIVATE AA PDP CONTEXT REQUEST message to the network.

As long as no auxiliary TLLI is allocated by the network to the mobile station, a random TLLI is used for addressing on the lower layers.

How the TLLI Is Transmitted

TLLI is not carried in an LLC frame or GMM message, but is carried in the RLC/MAC blocks on the U_m interfaces and in the BSSGP message on G_b interface. Figure 8-9 shows the format of the TLLI transfer:

- The TLLI is only transmitted on the RLC/MAC uplink data block when a GMM procedure is used (that is, when just the first blocks are sent during a signaling procedure). The *Transaction Identifier* (TI) bit indicates if a TLLI number is present in the frame.

- The TLLI is always transmitted in the BSSGP messages for all uplink and downlink packet data units and in many signaling procedures. The *Information Element Identity* (IEI) bytes indicate to the BSSG protocol what sort of information it should find in the following bytes delimited with the Length Indicator byte.

Figure 8-9

How the TLLI is transmitted.

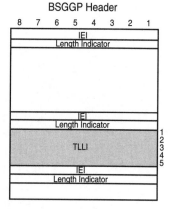

Mobility Management (MM)

The *Public Land Mobile Network* (PLMN) provides information for the mobile station to

- Detect when it has entered a new cell or a new RA.
- Determine when to perform periodic RA updates.

The mobile station detects when it has entered a new cell by comparing the cell's identity with the cell identity stored in the mobile station's MM context. The mobile station detects that a new RA has been entered by periodically comparing the *routing area identity* (RAI) stored in its MM context with that received from the new cell. The mobile station considers hysteresis in signal strength measurements.

When the mobile station camps on a new cell, possibly in a new RA, this indicates one of three possible scenarios:

- A cell update is required.
- An RA update is required.
- A combined RA and location area update is required.

In all three scenarios, the mobile station stores the cell identity in its MM context. If the mobile station enters a new PLMN, the mobile station either performs an RA update or enters the idle state.

A cell update takes place when the mobile station enters a new cell inside the current RA and the mobile station is in the ready state. If, however, the RA has changed, an RA update is executed instead of a cell update.

GPRS Attach Procedure

In the attach procedure, the mobile station provides its identity. The identity provided to the network is the mobile station's P_TMSI or IMSI:

- P_TMSI and the RAI associated with the P_TMSI is provided if the mobile station has a valid P_TMSI.
- If the mobile station does not have a valid P_TMSI, then the mobile station provides its IMSI.

At the RLC/MAC layer (through the air interface), the mobile station identifies itself with

■ A foreign TLLI if a valid P_TMSI is available
■ A random TLLI if a valid P_TMSI is not available

The foreign or random TLLI is used as an identifier during the attach procedure until a new P_TMSI is allocated. Note, though, that the mobile station identifies itself with a local or foreign TLLI if the mobile station is already GPRS-attached and is performing an IMSI attach.

At BSSGP layer (through the G_b interface), the mobile station identifies itself with the information field TLLI in its messages.

Upon receipt of the Attach Request message, SGSN allocates a new P_TMSI in this RAI, and transmits it to the mobile station in an Attach Accept (new P_TMSI) message. If P_TMSI was changed, the mobile station acknowledges the received P_TMSI with the Attach Complete (P_TMSI) message.

Cell Update in Packet Idle Mode

Normally, the mobile station stays in the cell selection and reselection mode. The cell update process occurs when the mobile station is in the ready state and in packet idle mode. The mobile station continually monitors the signal quality and strength from the adjacent cells and determines the power control of the received signal. If the mobile determines that a different cell has a better quality signal, then a cell reselection occurs. This occurs in the sequence that follows as a preferred way of performing the cell update:

1. The mobile station has selected a new cell in the current RA and is camping on it. No handover occurs in GPRS, but the network knows precisely where the mobile station is (if it is the ready state).

2. The mobile station performs the cell update procedure by sending an uplink LLC frame of any type containing the mobile station's identity to the SGSN. In the direction toward the SGSN, the BSS shall add the *Cell Global Identity* (CGI), including the *routing area code* (RAC) and *location area code* (LAC) to all BSSGP frames.

3. The SGSN records this mobile station's change of cell, and further traffic directed toward the mobile station is conveyed over the new cell (SGSN considers the new CGI).

Cell Update in the Packet Transfer Mode

Again, the mobile station normally stays in the cell selection and reselection mode. The mobile station continually monitors the signal quality and strength from the adjacent cells and determines the power control of the received signal. If the mobile determines that a different cell has a better quality signal, then a cell reselection occurs. The cell reselection process occurs with the mobile station in the ready state and in packet transfer mode as follows:

1. The mobile station is in packet transfer mode. As it gets near the limits of the cell, the radio quality decreases; therefore, the RLC/MAC layer asks for many retransmissions. As the radio quality decreases, the actual data reception also decreases. The LLC retransmissions increase.

2. The LLC layer begins to receive fewer packets. Now the mobile station, which is constantly in reselection mode, realizes that a new cell has a stronger C2+ cell_reselect_hysteresis.

3. Therefore, it releases its *Temporary Block Flow* (TBF) in the current cell, selects the new cell, and reestablishes a TBF in this new cell.

4. Now the LLC starts receiving packets again. If some of them were lost, LLC will request retransmissions. The previous process assumes that the RLC and LLC layers are in acknowledged mode.

Routing Area Updates (Intra-SGSN)

The RA update on an intra-SGSN arrangement is fairly simple and easy to accommodate. An RA update procedure occurs when the mobile station detects that it has entered a new RA or the periodic RA update timer has expired. The mobile station performs the RA update procedure as follows:

1. The mobile station sends an RA Update Request (old RAI and the update type) to the SGSN. The update type can be a periodic RA update or normal RA update.

2. This is preceded by an access request (one- or two-phase access with an MM procedure).

3. The SGSN allocates a new P_TMSI and sends it in the RA UPDATE ACCEPT message.

RA Updates (Inter-SGSN)

The need to perform an RA update when the mobile station enters a new SGSN area creates a more complex update procedure. This is not to imply that the update procedure is at risk or constantly fails, but just that the procedure requires more steps and takes on more complex decisions. Figure 8-10 shows the sequence.

1. The mobile station sends an RA Update Request (old RAI and the update type) to the SGSN. The update type can be either a periodic RA update or normal RA update. The PCU adds the CGI before passing the message to the SGSN.

2. The new SGSN sends an SGSN Context Request message (old RAI, TLLI, New SGSN Address) to the old SGSN to get the MM and PDP contexts for the mobile station. The old SGSN responds with an SGSN Context Response message (MM Context, PDP Contexts, LLC ACK). The LLC acknowledge contains the acknowledgments for each LLC connection used by the mobile station.

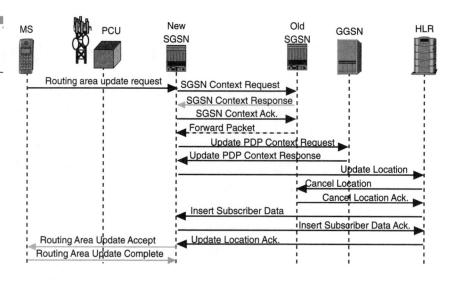

Figure 8-10
The RA update inter-SGSN.

3. The new SGSN sends an SGSN Context Acknowledge message to the old SGSN. The new SGSN is ready to receive data packets belonging to the activated PDP contexts.

4. The old SGSN duplicates the buffered *network layer protocol data units* (N-PDUs) and starts tunneling them to the new SGSN.

5. The new SGSN sends an Update PDP Context Request message (new SGSN Address, *tunnel identifier* [TID], QoS Negotiated) to the *Gateway GPRS Support Node* (GGSN) concerned. The GGSNs update their PDP context fields and return an Update PDP Context Response message (TID).

6. The new SGSN informs the *Home Location Register* (HLR) of the change of SGSN by sending an Update Location message (SGSN Number, SGSN Address, IMSI) to the HLR.

7. The HLR sends a Cancel Location message (IMSI, Cancellation Type) to the old SGSN with a Cancellation Type set to Update Procedure message. The old SGSN acknowledges with a Cancel Location ACK message (IMSI).

8. The HLR sends an Insert Subscriber Data message (IMSI, GPRS subscription data) to the new SGSN. If all checks are successful, then the SGSN constructs an MM context for the mobile station and returns an Insert Subscriber Data ACK message (IMSI) to the HLR. The HLR acknowledges by sending an Update Location ACK message (IMSI) to the new SGSN.

9. The SGSN allocates a new P_TMSI and sends it in the RA Update Accept (new P_TMSI, LLC ACK) message.

10. The mobile station acknowledges the new P_TMSI with an RA Update Complete (P_TMSI, LLC ACK) message.

SNDCP Layer

Network layer protocols are intended to operate over services derived from a wide variety of subnetworks and data links. GPRS supports several network layer protocols providing protocol transparency for users of the service. The introduction of new network layer protocols will therefore be possible without changing any of the lower-layer GPRS protocols. Therefore, all functions related to the transfer of N-PDUs are carried out transparently by GPRS network entities. The SNDCP is situated below the network layer and above the LLC layer. It provides protocol transparency as it supports a variety of network layer protocols.

The set of protocol entities sitting above SNDCP consists of commonly used network protocols. These all use the same SNDCP entity, which performs the multiplexing of data coming from the different sources before being sent via the services provided by the LLC layer. The NSAPI acts as an index for the appropriate PDP that is using the services of SNDCP. Each active NSAPI uses the services provided by the SAPI in the LLC layer and as such several NSAPIs may be associated with the same SAPI.

The SNDCP supports compression of redundant user data and protocol control information. Data compression could be V.42-bis compression, plus an optional TCP/IP header compression. Several network layer applications using the same QoS class (or the same SAPI) could be compressed using one compressor. Compression is one of the optional functions of the SNDCP layer. The network layer protocols share the same SNDCP that then performs multiplexing of the data coming in from the different sources to be sent across the LLC to the mobile station.

The outputs from the compression subfunctions are segmented to maximum length LLC frames before sending them over the LLC. On the other side, the SNDCP layer reassembles them before decompression. The SNDCP provides for the transmission and reception of N-PDUs in the acknowledged as well as the unacknowledged mode. In the acknowledged mode, the receipt of data is confirmed at the LLC layer and data transmission and reception is done in order. In the unacknowledged mode of transmission, the receipt of data is not confirmed.

SNDCP Identities

The SNDCP maps the network protocols to best fit the underlying GPRS transmission capabilities, as shown in Figure 8-11. It deals with the upper- and lower-layer primitives, and supports

- Network protocol packets priority management (a long file transfer may be bypassed by a short message of higher priority). SNDCP provides functions that help to improve channel efficiency. For this requirement, SNDCP uses compression technique.
- Segmentation if necessary to best fit the air resources.
- Sharing of a single low-layer connection by multiple network layers (that is, address management or priorities).

Figure 8-11
The SNDCP layer.

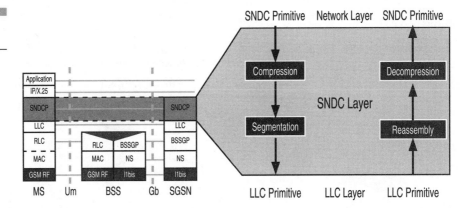

SNDCP Service Functions

The following functions are performed by the SNDCP and are shown in Figure 8-12:

- Transmission and reception of N-PDUs in acknowledged or unacknowledged LLC mode. In acknowledged mode, the receipt of data is confirmed at the LLC layer and the data is transmitted and received in order per NSAPI. In unacknowledged mode, the receipt of data is not confirmed at the SNDCP layer or the LLC layer.

- Transmission and reception of variable-length N-PDUs between the mobile station and the SGSN.

- Transmission and reception of N-PDUs between the SGSN and mobile station according to the negotiated QoS profile.

- Segmentation and reassembly of the data to a maximum-length PDU. The output of the compression function is segmented to the length of the PDU as determined by the mode selected. This is independent of the particular network layer protocol being used.

- Transfer of the minimum amount of data possible between the SGSN and the mobile station through the use of compression techniques.

- Compression of redundant protocol information (TCP/IP headers) at the transmitter and decompression at the receiver. Compression may be performed independently for each QoS delay class and precedence class. If several network layers use the same QoS delay class and

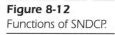

Figure 8-12
Functions of SNDCP.

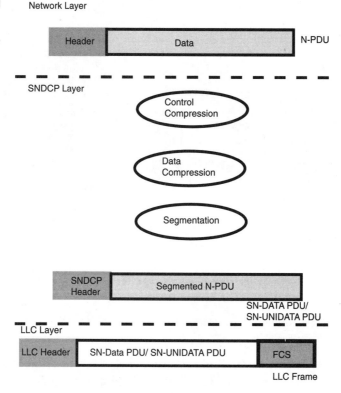

precedence class, then one common compressor may be used for these network layers.

SNDCP Layer—NSAPIs

The SNDCP layer takes the packet data information that comes from the network layer, known as the N-PDUs and adds a header that contains the NSAPI information onto it. Figure 8-13 shows the multiple layer protocols. These newly formed packets are called *subnetwork packet data units* (SPDUs). These are used to differentiate one network layer application from another. This identifier is unique for a particular network layer application for the duration of that data call.

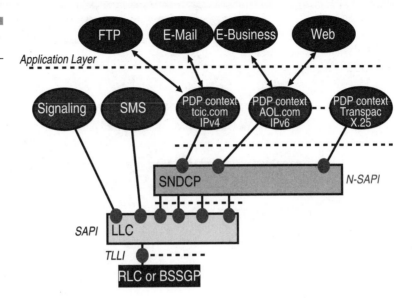

Figure 8-13
SNDCP NSAPIs.

A variety of network layers are supported (such as IP and X.25) above SNDCP. The network layer packet data protocols share the same SNDCP that performs multiplexing of data coming from the different sources to be sent across the LLC. To identify the different network layers (IP and X.25), the service needs to use the NSAPI. Several NSAPIs may be associated with one SAPI (they may use the same QoS profile). Several applications can use the same PDP context (for example, Netscape and e-mail).

SNDCP Compression and Segmentation

Separate data compression entities are used for acknowledged (SN DATA) and unacknowledged (SN UNIT DATA) data transfer.

SNDCP uses two types of compression, as seen in Figure 8-14:

■ Data header compression (TCP/IP header compression; only the differences between two consecutive headers are transmitted)

■ Data compression (V.42 bis)

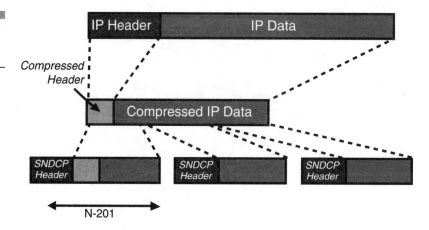

Figure 8-14
SNDCP compression
and segmentation.

The segmentation and reassembly procedures are different for acknowledged and unacknowledged mode of operation. Any (possibly compressed) combined N-PDU and SNDCP header is segmented by SNDCP if it is longer than the maximum number of bytes in the information field of an LLC frame:

- Maximum number of N201-I (1,520 bytes) is used for acknowledged data transfer mode (SN DATA segmentation).
- Maximum number of N201-U (500 bytes) is used for unacknowledged data transfer mode (SN UNIT DATA segmentation).
- The MORE (M) bit is used to indicate the last segment.

Two types of PDU formats are used by SNDCP, depending on the transfer mode. These are shown in Figure 8-15.

- In acknowledged mode, the SNDCP header is two bytes long.
- In unacknowledged mode, the SNDCP header is five bytes long.

Using the interfaces defined in this chapter has covered the various ways the Layer 3 protocols work with the upper-layer primitives of the Layer 2 protocols to perform the necessary truncation of the data packets so that they will work in the LLC format. The use of SNDCP NSAPIs passed down to the Layer 2 protocols helps to unambiguously identify the mobile station using the GPRS network. Through this interface, the GPRS network nodes

Figure 8-15
The SNDCP PDUs.

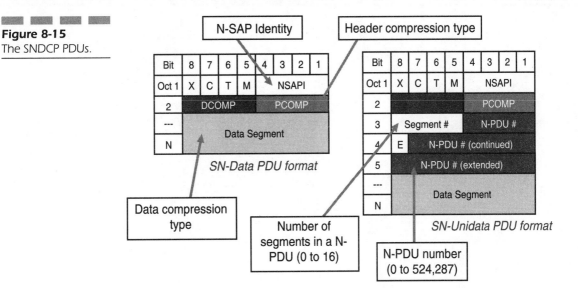

can isolate individual data flows when multiple mobile stations' input are multiplexed together on a single channel. These protocols were carefully planned to enable the interoperation among the various nodes in a GSM and GPRS architecture.

PCUSN-to-SGSN Interface (G_b)

Objectives

Upon completion of this chapter, you should be able to

- Describe the PCUSN interface.
- Discuss the way that Frame Relay works.
- Understand the way GPRS uses the Frame Relay services at the G_b interface.
- Combine the benefits of Frame Relay and the advantages of GPRS.
- Understand why ETSI chose Frame Relay as the transport for the PCUSN.

High-Level Characteristics of the G_b Interface

In contrast to the A interface, where a single user has sole use of a dedicated physical resource throughout the lifetime of a call irrespective of information flow, the G_b interface enables many users to be multiplexed over a common physical resource. GPRS signaling and user data may be sent on the same physical resources. Access rates per user may vary from zero data to the maximum possible bandwidth (for example, the available bit rate of an E1).

Position of BSSGP Within the Protocol Stack on the G_b Interface

The following peer protocols have been identified across the G_b interface: the *Base Station Subsystem GPRS Protocol* (BSSGP) and the underlying *network service* (NS). The NS transports BSSGP *packet data units* (PDUs) between a BSS and a *Serving GPRS Support Node* (SGSN). The primary functions of the BSSGP include

- In the downlink, the provision by an SGSN to a BSS of radio-related information used by the *Radio Link Control/Medium Access Control* (RLC/MAC) function

- In the uplink, the provision by a BSS to an SGSN of radio-related information derived from the RLC/MAC function
- The provision of functionality to enable two physically distinct nodes, an SGSN and a BSS, to operate node management control functions

The Protocol Stack for G_b Interface

The G_b interface is located between the *packet control unit* (PCU) and SGSN, as shown in Figure 9-1. On each node, we retrieve the following:

- **Physical layer**   This defines the characteristics of the medium being used.
- **NS**   This defines the Layer 2 protocol that is being used (Frame Relay will be used initially) and some specific procedures.
- **BSSGP**   This mainly manages buffers for flow control between the PCU and SGSN. It provides services for the upper-layer entities.
- *Network Management* **(NM)**   This local entity manages the buffers and virtual circuits between the two nodes.
  - *GPRS Mobility Management* **(GMM)**   The GMM part that is located just above the BSSGP part deals only with mobility messages between SGSN and PCU (such as a paging procedure).

Figure 9-1
The G_b protocol stack.

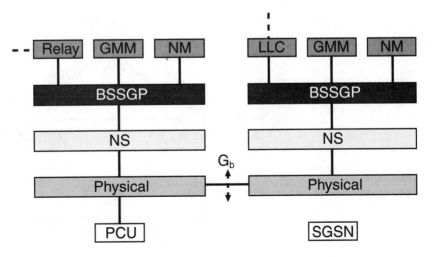

- *Logical Link Control* **(LLC) relay toward RLC/MAC**   These are the layers that are used for relaying packets through the PCU-SGSN points. For example, users' packets take these access points.

Frame Relay Networks

Frame Relay was selected by the *European Telecommunications Standards Institute* (ETSI) committees because of its robustness, its speed that operates at up to 2.0 Mbps, and the size of the frame set at up to 1,610 bytes in the payload. These coincide with the demands for GPRS. A Frame Relay network is a packet data network made of Frame Relay switches that are connected to each other, as shown in Figure 9-2. A user (user is a generic name, it could be a *Local Area Network* [LAN], for example) can gain access to the network by using a *Frame Relay Access Device* (FRAD). Its job is to build Frame Relay frames. The interface between FRAD and the Frame Relay network is well defined; this is the *User-to-Network Interface* (UNI).

The FRAD

A FRAD builds the frames that are used on the network, as shown in Figure 9-3. It is an interworking device between the user and the Frame Relay

Figure 9-2
A Frame Relay network.

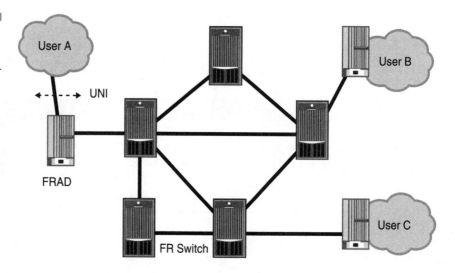

Figure 9-3
The FRAD.

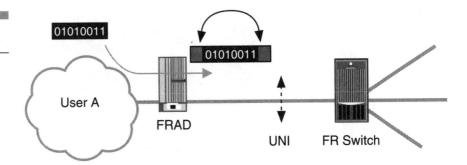

network. It can be a router connected to a LAN, for example, or end-user equipment.

The Protocol Data Unit (PDU)

The basic frame structure for other synchronous protocols is the same as most *high-level data link control* (HDLC) frame formats. In this case, the HDLC frame header consists of an address and control information. For Frame Relay, the header is changed, as shown in Figure 9-4, and uses the 2 bytes (octets) to define the following pieces:

■ *Data Link Connection Identifier* **(DLCI)** This identifier uses a 10-bit field and supports up to 1,024 *logical connection numbers* (LCNs).

■ *Command/Response* **(C/R) bit** The C/R is a standard bit used in HDLC framing (an OSI standard). This is not used in Frame Relay.

■ *Extended Address* **(EA) bit** When set to 0, this extends the DLCI address. When set to 1, this indicates that the address is only carried in the first byte (6 bits).

■ *Forward Explicit Congestion Notification* **(FECN)** This is set in the frames going out into the network toward the destination address.

■ *Backward Explicit Congestion Notification* **(BECN)** This is set in frames returning from the network to the source address.

■ *Discard Eligibility* **(DE) bit** This bit is used by the source equipment to denote whether the frame is eligible to be discarded by the network if the network gets congested. When set to 1, this indicates that the frame is eligible to be discarded during congestion period.

■ **EA bit** When set to 1, this is used to end the DLCI.

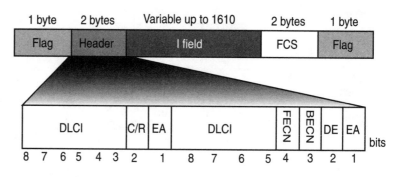

Figure 9-4
The frame PDU.

Where:
BECN   Backward Explicit Congestion Notification
C/R   Command/Responses
DE   Discard Eligibility
DLCI   Data Link Connection Identifier
EA   Extended Address
FECN   Forward Explicit Congestion Notification
PDU   Protocol Data Unit

The FECN, BECN, and DE parts of the addresses are designed to alert the end-user equipment on the status of the network's capability to process frames. Higher-level protocols use a window size control procedure or other form of control to handle traffic flow. These flow control mechanisms are designed to alert the network components and the end-user equipment to slow the delivery of frames to the network. If discarding is taking place while a router or other piece of equipment is in burst mode, the router or other equipment will begin to buffer the frames until the congestion is resolved.

Frame Relaying

Frame Relay networks use DLCIs to route the frames between two adjacent nodes. A DLCI identifies a channel.

Frame Relay switches have routing tables that associate an input port and a DLCI to an output port and another DLCI. Each frame that is received is forwarded to the correct port with a new DLCI. The data populating the tables points to the type of circuit being used such as

■ *Permanent Virtual Circuit* (**PVC**)   Used between two users, it consists of data that fills these routing tables; the frames then have the equivalent of a dedicated path through the network.

■ *Switched Virtual Circuit* (**SVC**)   This may also be established at will by the user on a per-call basis. It is not supported in GPRS in which only PVCs are used.

Benefits of Frame Relay

Frame Relay saves money on *customer premises equipment* (CPE), local access, and interexchange network. It also reduces administrative and operations costs. More direct connectivity between locations can be provisioned for a minimal incremental cost. Because of this, Frame Relay networks can be designed to better match the underlying traffic patterns.

Locations and connections can be added easily and more cost effectively with Frame Relay compared with private lines. It is easy to redesign and optimize the network because changing port connection and virtual circuit bandwidth is software configurable. Most Frame Relay networks have self-healing or automatic-rerouting capabilities between the Frame Relay switches.

Frame Relay is based on statistical multiplexing where bandwidth can be shared between active applications and/or connections only. This lowers the cost of ownership because end users don't have to pay for idle or excess capacity needed to meet peak traffic periods. In a branch office network environment, these cost savings can be substantial. Frame Relay can support multiple protocols and applications including LAN, *Systems Network Architecture* (SNA), on-net voice, and packetized video. End users can migrate from multiple parallel networks deployed to support each application to a single Frame Relay network. Simplification has many benefits including reducing costs, improving reliability and performance, and simplifying planning and reengineering.

Frame Relay is enhanced by the capability to smoothly migrate to *Asynchronous Transfer Mode* (ATM) when higher capacity is required; this enables customers to get started with the more complex ATM in a gradual fashion and to learn the complexities and then apply them to the network as needed.

Service Comparison

Private lines use *Time Division Multiplexing* (TDM) where fixed time slots or channels are dedicated to specific applications whether the applications are active or not. Private lines are generally designed to meet peak traffic delivery objectives. This translates into excess bandwidth or idle capacity during non-peak hours. Private lines are more susceptible to network downtime because the network does not automatically reroute around physical network failures, such as cable cuts, unless the end user invests in intelligent network switches and geographically diverse facilities for backup purposes.

Statistical multiplexing with Frame Relay enables applications to share network bandwidth. Network bandwidth is used by active applications only. An active application can have full access to the port connection for the entire duration of time when other applications that share that port are not transmitting or receiving. Frame Relay can automatically reroute around network failures because the Frame Relay switches have the intelligence and capability to obtain network status, interpret the status, and take the appropriate action depending on the interpretation, like redirecting traffic to a different trunk because the primary trunk has failed.

Why Frame Relay Was Developed

Major trends in the industry led to the development of Frame Relay services, as shown in Figure 9-5. These can be categorized into four major trends:

■ **The increased need for speed across the network platforms within the end-user and the carrier networks** The need for higher speeds is driven by the move away from the original text-based services to the current graphics-oriented services and the bursty, time-sensitive data needs of the user through new applications. The proliferation of LANs and now the client/server architectures that are being deployed have shifted the paradigm of computing platforms. The demands of these services will exceed the data transport needs of the older text-based services by hundreds of thousands of times. Users demand more readily available connectivity and the speed to ensure quick and reliable communications between systems or services. Fortunately, the bursty nature of the way we conduct our business

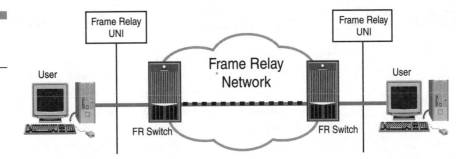

Figure 9-5
Why Frame Relay
was developed.

enables the sharing of resources among many users who thereby share
the bandwidth available. To accommodate this connectivity in a quick
manner, some changes had to be made, and the protocol dependency
and processing of the networks had to be minimized. One way to
accommodate the reduced overhead associated with the network was
to eliminate some of the processing, mainly in the error detection and
correction schemes.

- **Increasing intelligence of the devices attached to the network**
  The use of data transfer between and among devices on the network
  has moved many of the processing functions to the desktop. Because
  the processing is now being conducted at the local device, as opposed to
  using dumb terminals and a single host processor, the capacity to move
  the information around the network must meet the demands of each
  attached device. Increased functionality must be met with increases in
  the bandwidth allocation for these devices.

- **Improved transmission facilities** The days of dirty or poor-quality
  transmission lines required the use of overcorrecting protocols such as
  X.25 and SNA. Because the network now performs better, a newer
  transmission capability is needed.

- **The need to connect LANs to *Wide Area Networks* (WANs) and
  the internetworking capabilities** Today's users want to connect
  LANs across the boundaries of the wide area, unshackling themselves
  from the bounds of the LAN. The users demand and expect the same
  speed and accuracy across the WAN that they have on the local
  networks. Therefore, a new transport system to support the higher-
  speed connections across a wider area was needed. The LAN-to-WAN
  internetworking works fine in a simple *point-to-point* (PTP)
  arrangement, except that the network is dynamic, and the ability to
  connect to multiple sites concurrently must be robust enough to meet
  this new need.

The Significance of Frame Relay

The network was originally brought up through the older analog transmission techniques, which have been addressed several times. As an analog transmission system, the network was extremely noisy and produced a significant amount of network errors and data corruption. This element was most frustrating to the data-processing departments. When data errors were introduced, a retransmission was required. The more retransmissions were necessary, the less effective the throughput was on the network. In fact, several years ago, the use of a 4,800-bps-transmission service on the analog dial-up network might have produced an effective throughput of only 400 bps after all the errors and retransmissions. This was intolerable and had to be corrected. To solve this problem, the network introduced the X.25 services, also called packet switching. Whereas the X.25 was originally designed to handle the customer's asynchronous traffic, Frame Relay was designed to take advantage of the network's capability to transport data on a low-error, high-performance digital network and to meet the needs of the intelligent, synchronous use of the newer, more sophisticated user applications. The protocols that apply to the basic needs of the data transmission were used, yet much of the overhead has been squeezed out, creating a service that operates functionally at the bottom half of Layer 2 on the OSI model, as shown in Figure 9-6.

When compared to private leased lines, Frame Relay makes the design of a network much simpler. A private-line network requires a detailed analysis to set all the right connections in place; this further accentuates the traffic-sensitive needs of the user network. The meshed network uses a

Figure 9-6
The Frame Relay protocols compared to the OSI model.

Typical OSI Stack

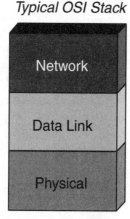

Network

Data Link

Physical

Frame Relay Stack

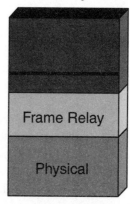

Frame Relay

Physical

series of connections that are the total number of sites less one divided by two (n × [N − 1] / 2) points with links running from site to site, as shown in Figure 9-7. Therefore, if 10 sites exist in the network, nine local links will run from each site to every other site (10 × (10 − 1) / 2 = 45). This enhances the speed of connectivity, but the network costs are much higher. Further, depending on the nature of the data traffic, the bursty data needs of a LAN-to-LAN or LAN-to-WAN connection are not required full time. We, therefore, spend significantly more of the organization's money to support the meshed leased-line network.

A Frame Relay access from each site is provided into the network cloud, requiring only a single connection point rather than the nine of the earlier network, as shown in Figure 9-8. Data transported across the network will be interleaved on a frame-by-frame basis. Multiple sessions can run on the same link concurrently. Communications from a single site to any of the other sites can be easily accommodated using the predefined network connections of the virtual circuits. In Frame Relay, these connections use *Permanent Logical Links* (PLLs), more commonly referred to as PVCs. Each of the PVCs connects two sites just as a private line would, but in this case, the bandwidth is shared among multiple users rather than being dedicated to the one site for access to a single site. Using this multiple-site connectivity on a single link reduces the costs associated with CPE, such as CPU ports, router ports, or other connectivity arrangements. Because fewer ports

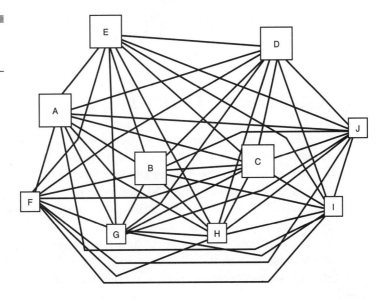

Figure 9-7
The meshed
network.

Figure 9-8
The Frame Relay
solution.

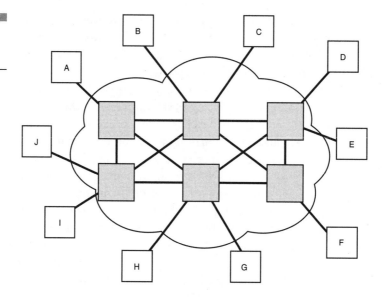

are required, fewer connection devices are required; therefore, the customer (in this case, the GPRS operator) saves money.

Because the PVCs are predefined for each in a pair of end-to-end connections, a network path is always available for the customer's application to run and transport data across the network. This eliminates the call setup time associated with the dial-up lines and the X.25 packet arrangements. The connection is always ready for the devices to ship data in a framed format as the need arises. This takes away the need for the constant fine-tuning of a private-line or a dial-up network link arrangement.

The Basic Data Flow

In most popular synchronous protocols, data is carried across a communications line based on very similar structures. The standard HDLC frame format is used in a myriad of these protocols and services. Frame Relay makes a very slight change to the basic frame structure, redefining the header at the beginning of the frame (2 bytes long). Figure 9-9 shows the basic data flow.

Although Frame Relay purports to eliminate the operations at the network layer, it does not eliminate all network layer operations. Figure 9-9 illustrates one network layer operation that is essential for Frame Relay

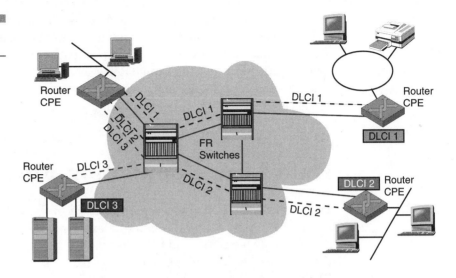

Figure 9-9
The basic data flow.

operations: the identification of virtual connections. Frame Relay uses the DLCI to identify the destination address. This 10-bit number corresponds to the virtual circuit number in the network layer protocol.

The DLCIs are premapped to a destination node, as shown in Figure 9-10. This simplifies the process at the routers because they only need to consult their routing table, check the DLCI in the table, and route the traffic to the proper output port based on this address.

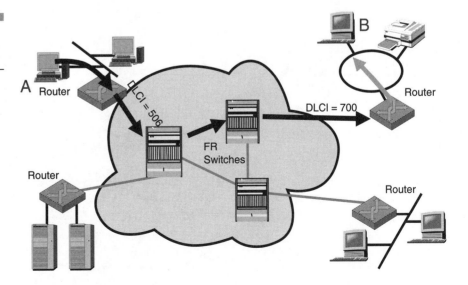

Figure 9-10
The mapping of the DLCI.

Inside the network, the same scheme is used although the Frame Relay switches need not maintain a strict virtual relationship in the network. Connectionless operations can be implemented to enable dynamic and robust routing between the Frame Relay switches. The only requirement is to make certain that the frame arrives sequentially at the port designated in the DLCI.

Data Link Connection Flow

The Frame Relay UNI enables multiple users to share the physical link. In Figure 9-11, users A and C are multiplexed onto the UNI by a router and the assigned DLCIs 499 and 506. The traffic is transported to the receiving Frame Relay switch, where it is presented to a router. Notice that the DLCIs are translated and mapped into DLCIs 700 and 530 at the remote UNI.

Because Frame Relay is a connection-oriented technology and uses labels DLCIs to identify traffic, a router must be able to translate a connectionless address to a DLCI and vice versa. Although this operation is not complex, it does require the careful construction of mapping tables at the router. The Frame Relay standards do not describe how this mapping and address translation takes place. Typically, each router has a table that correlates IP addresses to DLCIs and vice versa.

Figure 9-11
The translated mapping of the DLCI.

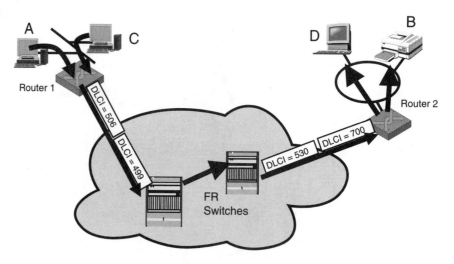

The FECN and BECN

The congestion and flow control option is optional; vendors need not implement this, and they will still comply with the standard. However, unless other flow control measures are implemented in the network, the use of this option is quite important.

Two mechanisms are employed to (a) notify users, routers, and Frame Relay switches about congestion, and (b) take corrective action. The BECN bit and the FECN bit achieve both capabilities. Figure 9-12 shows the FECN and BECN flow.

Assume that a Frame Relay switch is starting to experience congestion problems due to its buffers (queues becoming full and/or experiencing a problem with memory management). It may inform both the upstream nodes and the downstream nodes of the problem by the use of the FECN and BECN bits, respectively. The BECN bit is turned on in the frame and is sent downstream to notify (potentially) the source of the traffic that congestion exists at a switch. This would permit the source to flow control its traffic until the congestion problem is solved. In addition, the FECN bit could be set and placed in a frame and sent to the upstream nodes to inform them that congestion is occurring downstream. One might question why the FECN is used to notify upstream devices that congestion is occurring

Figure 9-12
The FECN and BECN.

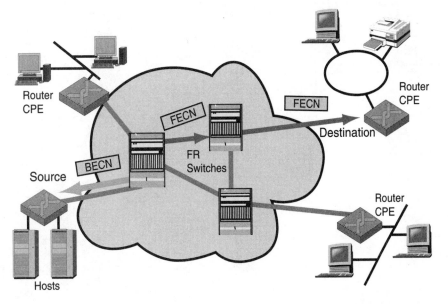

downstream. After all, the downstream device is the one creating the traffic problem.

The answer is that it varies, depending on remedial action at the upstream (destination) machines might want to take. For example, this FECN bit could be passed to an upper-layer protocol (such as the transport layer) that enables it to (a) slow down its acknowledgments (which in some protocols would close the transmit window at the destination device) or (b) establish its own more restrictive flow control agreement with its communications source machine (which also is permitted in some protocols). An obvious solution to the problem is for the source machine(s) to flow control itself to ameliorate the network congestion problem.

Frame Relay Speeds

Frame Relay was designed initially to start from 64 Kbps up to 1.544 Mbps in North America. Speeds of 2.048 Mbps were approved in the rest of the world. This speed is based on the use of T1 or E1 for the access link as specified for GPRS from ETSI. The Frame Relay network can be mapped to carry traffic from a larger location to multiple smaller locations, each operating at a different speed. Figure 9-13 shows the multiple flows from a single location on a high-speed circuit operating at 1.5 to 2.0 Mbps.

Figure 9-13
Multiple sites served by a single high-speed connection.

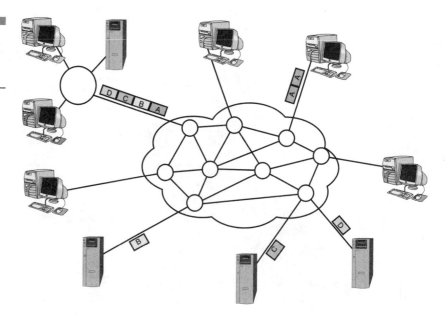

A small company came along and introduced speeds of 50 Mbps. This company, Cascade Communications, broke all the barriers. Cascade wanted the network to be robust, not limited to old data rates. Cascade was later acquired by Ascend Communications, and then Ascend was acquired by Lucent Technologies.

When designing a Frame Relay service, the speed of access is important both prior to and after installation. The customer must be aware of the need for and select a specified delivery rate. The speed can be assigned from both an access and a pricing perspective through various ways. For small locations, such as branch offices with little predictable traffic, the customer might consider the lowest possible access speed. The Frame Relay suppliers offer speeds that are flat rate, usage-sensitive, and flat-rate/usage-sensitive combined. The flat-rate service offers the speed of service at a fixed rate of speed, whereas the usage-based service might not include flat-rate service, but might have a pay-as-you-go rate for all usage. The combined service is a mix of both offerings. The customer selects a certain *committed information rate* (CIR). The CIR is a guaranteed rate of throughput when using Frame Relay. The CIR is assigned to each of the permanent virtual circuits selected by the user.

Each PVC is assigned a CIR that is consistent with the average expected volume of traffic to the destination port. Because Frame Relay is a duplex service (data can be transmitted in each direction simultaneously), a different CIR can be assigned in each direction. This produces an asymmetrical throughput based on demand. For example, a customer in Boston might use a 64-Kbps service between Boston and San Francisco for this connection; yet for the San Francisco-to-Boston PVC, a rate of 192 Kbps can be used. This provides added flexibility to meet the customer's needs for transport. However, because the nature of LANs is that of bursty traffic, the CIR can be burst over and above the fixed rate for two seconds at a time in some carriers' networks. This *committed burst rate* (Bc) is up to the access channel rate, but many of the carriers limit the burst rate to twice the speed of the CIR. When the network is not very busy, the customer could still burst data onto the network at an even higher rate. The *excess burst rate* (Be) can be an additional speed of up to the channel capacity, or in some carriers' networks, it can be 50 percent above the burst rate. Combining these rates, an example can be drawn as follows:

$$CIR + Bc + Be = Total\ throughput$$

Remember that the burst and the burst excess rates are for two seconds or less, depending on the carrier used. Some carriers do not allow any bursting across the network. Rather, they require that the maximum throughput be limited to the CIR. We are emphasizing that no standard offerings exist.

Provisioning PVCs and SVCs

The primary difference between PVCs and SVCs is whether the connections are provisioned or established. Both types of connections need to be defined. The difference is when the connections are defined and resources allocated.

The network operator typically provisions PVCs. The network operator can be the carrier (public services) or the *Management Information System Support* (MIS) manager (private networks). Once the PVC is provisioned, the connection is available for use at all times unless a service outage occurs. On the other hand, the end user, not the network operator, establishes SVCs. Prior to each use, an SVC is established to the destination end user. The connection is cleared after use.

SVC UNIs and NNIs

The Frame Relay Forum has two implementation agreements specifying the implementation of SVCs over UNIs and *Network-to-Network Interfaces* (NNIs).

The SVC UNI agreement depicts how an SVC can be established and released from an end-user device. On the other hand, the SVC NNI agreement depicts how an SVC can be established and released between two or more independent Frame Relay networks. The interconnected Frame Relay networks can both be private, public, or one public and one private.

The Network-to-Network Interface (NNI)

The initial thrust of the Frame Relay work focused on the UNI. Subsequent work resulted in the publication by the Frame Relay Forum (based on ANSI's T1.617 Annex D) of a NNI. This interface is considered instrumental to the success of Frame Relay because it defines the procedures for different networks to interconnect with each other to support the Frame Relay operations.

Obviously, the UNI defines the procedures between the user and the Frame Relay network, and the NNI defines the procedures between the Frame Relay networks. A PVC operating across more than one network is called a multinetwork PVC. Each piece of the PVC provided by each network is a PVC segment. Therefore, the multinetwork PVC is the combination of the relevant PVC segments. In addition, the NNI uses the bidirectional network procedures, published in ANSI T1.617 Annex D, and further requires that all networks involved in the PVC must support NNI procedures as well as UNI procedures.

Full internetworking operations between Frame Relay networks require that the procedures stipulated in ANSI T1.617 Annex D be used at the UNI and the NNI. This concept means that a user sends a *status enquiry* (SE) message to the network, and the network responds with a *status* (S) message. In addition, bidirectional procedures at the NNI require that either network be able to send SE and S messages.

Frame Relay/ATM Interworking

Two types of Frame Relay-to-ATM interworking are available: network and service. Service providers use network interworking to reduce congestion and achieve economies of scale in the backbone. End users are not impacted by such a deployment, as the protocol in and out of the cloud is still Frame Relay. Figures 9-14 and 9-15 show the use of service and network interworking.

Demand continues to increase for service interworking as some end users are beginning to see a need for ATM for some corporate locations. The

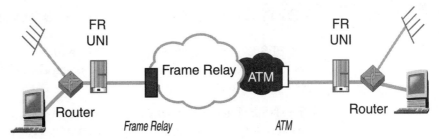

Figure 9-14
Service interworking.

Figure 9-15
ATM interworking.

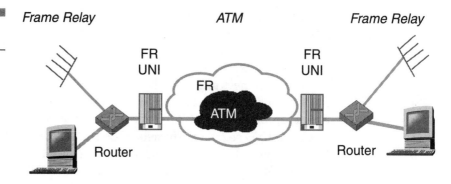

most typical example is the end user with a large number of remote locations running Frame Relay—all needing connectivity with the headquarters location. The headquarters supporting a large quantity of remote-site traffic may require higher bandwidth connections. Currently, the end users have three options: multiple T-1 Frame Relay ports, a high-speed Frame Relay port, or an ATM port. End users running voice, video, and data and having already invested in ATM customer premises equipment at the headquarters may benefit most from deploying ATM at this location. With service interworking, the network is responsible for protocol conversion, enabling a Frame Relay site to communicate with an ATM site and vice versa. ETSI is expected to support ATM in the future as the broader band data needs are satisfied.

In the interim, as the users and operators combine their needs, the possibility is also one that X.25 and Frame Relay interworking may also exist. Figure 9-16 shows this possibility as the network provider provides the interfaces through the routers. This is not an ETSI specification, but a hypothetical possibility because in many parts of the world, X.25 is still the primary data access mechanism from the carriers and end-user interfaces alike.

Network Service Sublayers

The GPRS Layer 2 interface for the SGSN and *Packet Control Unit Support Node* (PCUSN) is shown in Figure 9-17 using the NS layer, which is composed of two parts:

■ The NS Subnetwork part that defines the Layer 2 protocol that will be used (Frame Relay, ATM). In the beginning, Frame Relay will be used.

Figure 9-16
Frame and X.25
interworking.

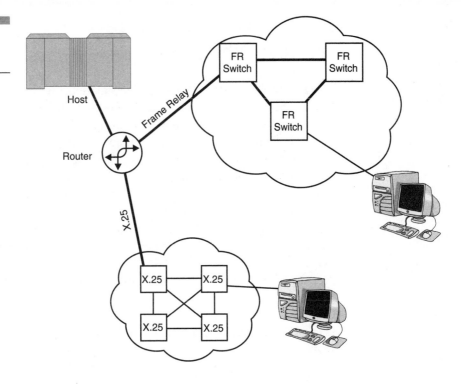

Figure 9-17
The NS layer.

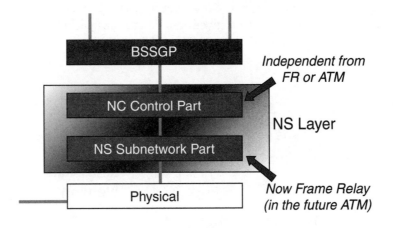

- The NS Control part creates circuit identifiers and defines procedures to manage them. Each NS virtual circuit that is created is associated to a Frame Relay virtual circuit.

Identifiers Managed by the NS Layer

The Frame Relay network defines a local identifier for a virtual circuit: the DLCI. The NS layer uses an end-to-end logical identifier to reference the same virtual circuit: the *Network Service Virtual Connection Identifier* (NSVCI). The purpose of the new identifier is to create software that will manage the NSVCI, not the DLCI. Thus, if the NS Subnetwork evolves (toward ATM, for example), the upper-layer coding will not be impacted. Figure 9-18 represents two PCUSNs connected to an SGSN using two different NSVCIs. The *Network Service Entity* (NSE) at the BSS and the SGSN provides the network management functionality required for the operation of the G_b interface. Each NSE is identified by means of a *Network Service Entity Identifier* (NSEI). The NSEI together with the *BSSGP Virtual Connection Identifier* (BVCI) uniquely identifies a BVC (for example, a PTP functional entity) within an SGSN. The NSEI is used by the BSS and the SGSN to determine the NSVCs that provide service to a BVCI.

An NSE manages a pool of NSVCIs toward a specific node, as shown in Figure 9-19:

- In a PCU, one NSE is present.
- In an SGSN, many NSEs are defined (one per PCU).

For an NSE, the NS layer performs load sharing between NSVCIs to face traffic, but for a given mobile station, packets will always take the same NSVCI (so that the order is guaranteed).

Figure 9-18
The NS layer identities.

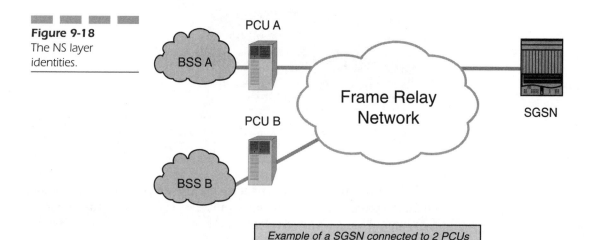

Example of a SGSN connected to 2 PCUs

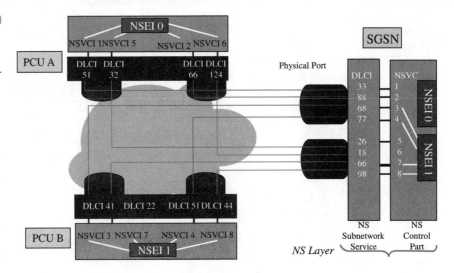

Figure 9-19
The use of the
identifiers for NSVCIs.

Network Service Control Procedures

The NS layer specification deals with giving logical identifiers to NSE (NSEI and NSVCI), but it also defines procedures between two NSE. Among them, we find the following:

- PDU transmission
- NSVC test
- NSVC reset
- NSVC blocking and unblocking

BSSGP Identifiers

BSSGP protocol functions satisfy many different conditions, which are shown in Figure 9-20. The BSSGP identifiers create the scenario to manage multiple entities and protocols:

- **As concerns LLC**
    - Transport of LLC packets through the G_b interface, LLC-UNITDATA messages (need to be sent in order).

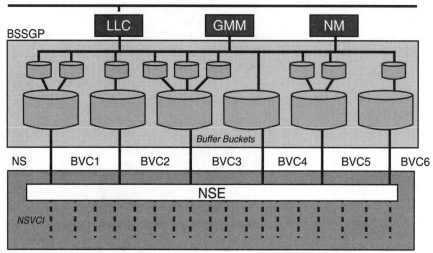

Figure 9-20
BSSGP identifiers.

- **As concerns NM**
  - Management of buffers at SGSN for flow control between PCU and SGSN (downlink)
  - LLC flushing and flow control messages
  - Supervision of *BSSGP Virtual Channels* (BVC). One BVC per cell is defined.
  - NM-status G_b Blocking/Unblocking, and G_b Reset messages.
- **As concerns GMM**
  - Paging messages (will be used when mobile terminating calls are supported)
  - Radio status message to report a failure linked to the radio interface

PDU Transmission

The BSSGP layer manages two kinds of buffers, as shown in Figure 9-21:

- **Mobile station buffers**   One for each user
- **BVC buffer**   One for each BVC

When receiving an LLC packet, BSSGP puts it into the mobile station buffer corresponding to the *Temporary Logical Link Identity* (TLLI). Mes-

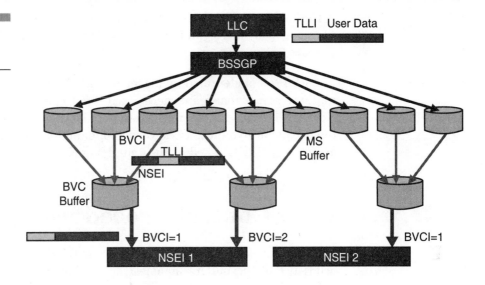

Figure 9-21
The PDU
transmission.

sages for the same cell are multiplexed onto the same BVC, and a buffer is
dedicated to it.

BSSGP Virtual Connection Identifier (BVCI)

BVCs provide communication paths between BSSGP entities. Each BVC is
used in the transport of BSSGP PDUs between peer PTP functional entities, peer *point-to-multipoint* (PTM) functional entities, and peer signaling
functional entities. The BVCI is used to enable the lower NS layer to efficiently route the BSSGP PDU to the peer entity. This parameter is not part
of the BSSGP PDU across the G_b interface, but is used by the NSE across
the G_b.

Any BSSGP PDU received by the BSS or the SGSN containing a PDU
type that does not fit (according to the definitions) with the functional
entity identified by the BVCI provided by the NSE is discarded and a STA-
TUS PDU with a cause value set to Protocol error—unspecified is sent.

- **PTP functional entity** Responsible for PTP user data transmission.
  One PTP functional entity exists per cell.

- **PTM functional entity** Responsible for PTM user data
  transmission. One or more PTM functional entities exist per BSS.

■ **Signaling functional entity** Responsible for other functions such as paging. Only one signaling entity exists per BSS.

■ **NSE** One or more NSEs exist per BSS.

Each BVC is identified by means of a BVCI, which has end-to-end significance across the G_b interface. Each BVCI is unique between two peer NSE. In the BSS, it is possible to configure BVCIs statically by administrative means or dynamically. In case of dynamic configuration, the BSSGP accepts any BVCI passed by the underlying NSE.

Flow Control Procedures

The PCU can send flow control messages to the SGSN to change the characteristics of the buffers managed by BSSGP (mobile station and BVC buffers) shown in Figure 9-22. That is to say

■ *Buffer maximum* (Bmax) (the size of the buffer − default value = 72 kilobytes) for the BVCI

■ *Rate of flow* (R) (the data rate of the leakage − default value = 10 Kbps) for the BVCI

■ Bmax (the size of the buffer − default value = 9 kilobytes) for each mobile station

■ R (the data rate of the leakage − default value = 0 bits) for each mobile station

Figure 9-22
The flow control procedure.

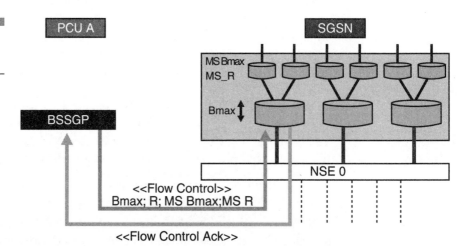

Each flow control message needs to be acknowledged. The procedure aims at adapting data rate on radio interface (RLC/MAC protocol) to the GPRS network.

Mode of Operation

The flow control mechanism manages the transfer of BSSGP UNITDATA PDUs sent by the SGSN on the G_b interface to the BSS. The BSS controls the flow of BSSGP UNITDATA PDUs to its BVC buffers by indicating to the SGSN the maximum accepted throughput in total for each BVC. The BSS controls the flow of BSSGP UNITDATA PDUs to the BVC buffer for an individual mobile station by indicating to the SGSN the maximum accepted throughput for a certain TLLI.

The BSS uses flow control to adjust the flow of BSSGP UNITDATA PDUs to a BVC buffer. The amount of buffered BSSGP UNITDATA PDUs in the BSS should be optimized to efficiently use the available radio resource. The volume of buffered BSSGP UNITDATA PDUs for a BVC or mobile station should be low. BSSGP UNITDATA PDUs queued within the BSS that are not transferred across the radio interface before the PDU life-time expires are locally deleted from the BSS. The local deletion of BSSGP UNITDATA PDUs in the BSS is signaled to the SGSN by the transmission of an LLC-DISCARDED PDU.

For each FLOW-CONTROL PDU received by an SGSN, a confirmation is always sent across the G_b interface by the SGSN. The confirmation uses the tag that was received in the FLOW-CONTROL PDU, which was set by the BSS to associate the response with the request. When receiving no confirmation to a FLOW-CONTROL PDU, the reasons that gave rise to the triggering of a flow control message may trigger another message, or, if the condition disappears, it may not. For the repetition of nonconfirmed FLOW-CONTROL PDUs, the maximum repetition rate still applies in the BSS.

Control of the Downlink Throughput by the SGSN

The principle of the BSSGP flow control procedures is that the BSS sends flow control parameters that enable the SGSN to locally control its transmission output in the SGSN to BSS direction. The SGSN performs flow control on each BVC and on each mobile station. The flow control is performed on each LLC-PDU first by the mobile station flow control mechanism and then by the BVC flow control mechanism. If the LLC-PDU is passed by the individual mobile station flow control, the SGSN then applies the BVC flow control to the LLC-PDU. If an LLC-PDU is passed by both flow control mechanisms, the entire LLC-PDU is delivered to the network services for transmission to the BSS.

The flow control parameters sent by the BSS to the SGSN consist of the following information:

- The bucket size (Bmax) for a given BVC or mobile station in the downlink direction

- The bucket leak rate (R) for a given BVC or mobile station in the downlink direction

- The bucket full ratio for a given BVC or mobile station in the downlink direction, if the *current bucket level* (CBL) feature is negotiated

The SGSN performs flow control on an individual mobile station using SGSN determined values of Bmax and R unless it receives a FLOW-CONTROL-MS message from the BSS regarding that mobile station. The SGSN continues to perform flow control for a particular mobile station using the Bmax and R values received from the BSS for at least T(h) seconds after receiving a FLOW-CONTROL-MS message from the BSS regarding that mobile station. When timer T(h) has expired or when the mobile station changes cells, the SGSN may reinitialize the SGSN internal flow control variables for that mobile station and begin to use SGSN generated values for Bmax and R.

Currently, the use of Frame Relay makes sense because the GPRS networks operate at speeds of up to 2 Mbps. In the future as the per-user speeds increase (such as with EDGE or 3G), the use of ATM will most likely be required. The greater the speed per mobile station, the greater the data-carrying capacity of the network to support the multitude of simultaneous users. ETSI was very clear that they intended this evolution to be a consideration in their decision to select the various protocols. In fact, many vendors' SGSN products may be a form of ATM switch already in preparation for this evolution.

One thing is certain: low-latency, high-speed, and highly reliable data transfer is a requirement to meet the demands of a mobile user. Frame Relay brings a good portion of the requirements to the table in its existing state. This will satisfy the network for the near term, or may be an interworking function at the PCUSN to SGSN that is converted to ATM in other stages of transition across the network.

10

SGSN-to-GGSN (G_n) and GGSN-to-PDN (G_i) Interface

Objectives

Upon completion of this chapter, you should be able to

- Describe the SGSN-to-GGSN interface.
- Discuss the GPRS Tunneling Protocols.
- Understand the way GPRS provides data security across the PLMN.
- Describe the components that can assist in securing the data.
- Understand why ETSI chose the use of IPSec and other Layer 2 protocols.

GPRS Tunneling Protocol (GTP)

The *GPRS Tunneling Protocol* (GTP) is the protocol between *GPRS Support Nodes* (GSNs) in the *Universal Mobile Telephone Systems / General Packet Radio Systems* (UMTS/GPRS) backbone network. It includes both the GTP signaling and control (GTP-C) and user data transfer (GTP-U) procedures. The two different types of tunnels deal with either network signaling and control for control purposes and for actual user data.

GTP is defined for the G_n interface, the interface between GSNs within a *Public Land Mobile Network* (PLMN), and for the G_p interface, the interface between GSNs in different PLMNs. Only GTP-U is defined for the Iu interface between the *Serving GPRS Support Node* (SGSN) and the *UMTS Terrestrial Radio Access Network* (UTRAN). On the Iu interface, the *Radio Access Network Application Part* (RANAP) protocol is performing the control function for GTP-U. Figure 10-1 shows the reference model for the GTP.

GTP enables multiprotocol packets to be tunneled through the UMTS/GPRS backbone between GSNs and between SGSN and UTRAN. In the signaling plane, GTP specifies a tunnel control and management protocol (GTP-C), which enables the SGSN to provide *packet data network* (PDN) access for a mobile system. Signaling is used to create, modify, and delete tunnels.

In the transmission plane, GTP uses a tunneling mechanism (GTP-U) to provide a service for carrying user data packets. The choice of path depends on whether or not the user data that will be tunneled requires a reliable link.

The GTP-U protocol is implemented by SGSNs and *Gateway GPRS Support Nodes* (GGSNs) in the UMTS/GPRS backbone and by *Radio Network*

Figure 10-1
The GTP reference
model.

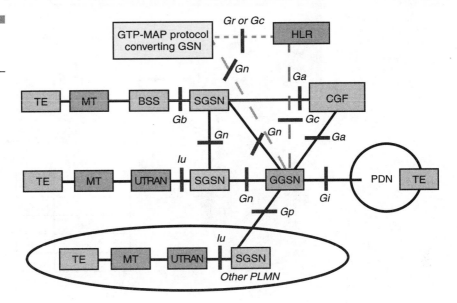

Controllers (RNCs) in the UTRAN. SGSNs and GGSNs in the UMTS/GPRS
backbone implement the GTP-C protocol.

GTP is the protocol used between GSNs, as shown in Figure 10-2. As the
GGSN may be linked to different kinds of PDNs, GTP enables multiprotocol packets to be tunneled through the GPRS backbone on the G$_n$ interface
(between GSNs within a PLMN) and on the G$_p$ interface (between GSNs in
different PLMNs).

GTP tunnels utilize the *Transmission Control Protocol/Internet Protocol*
(TCP/IP) for protocols that need a reliable data link (such as X.25) and the
User Datagram Protocol/Internet Protocol (UDP/IP) for protocols that do
not need a reliable data link (such as IP). GTP includes the signaling and
data procedures.

In the signaling plane, GTP specifies a tunnel control and management
protocol that enables the SGSN to provide GPRS network access for a
mobile station. Signaling is used to create, modify, and delete tunnels.

In the transmission plane, GTP uses a tunneling mechanism to provide
a service for carrying user data packets.

- Signaling plane
    - Path management messages (Echo Request/Echo Response
      messages)
    - Tunnel management messages

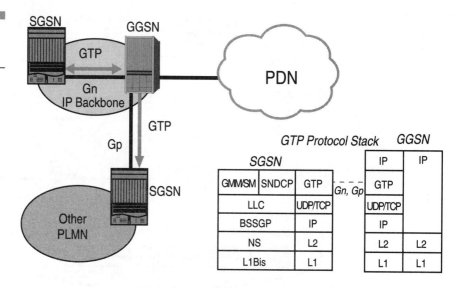

Figure 10-2
The GPRS Tunneling Protocol (GTP).

- Location management messages
- *Mobility management* (MM) messages
- Transmission plane
  - Tunnels are used to carry encapsulated *tunneled PDUs* (T-PDUs) between a given GSN pair (GGSN-SGSN or SGSN-SGSN) for individual mobile stations.
  - The key tunnel ID, present in the GTP header, indicates to which tunnel a particular T-PDU belongs.

GTP Messages

The GTP header is a fixed-format, 20-octet header used for all GTP messages, as shown in Figure 10-3.

- **Version**   Is set to 0 to indicate that this is the first version of GTP.
- **Spare 1111**   Are unused bits, set to 1 by the sending side, and not evaluated by the receiving side.
- **Message type**   Indicates the type of message, is set to the decimal value 255 for T-PDU, and takes a value from 1 to 52 for signaling messages.
- **Length**   Gives the size of the GTP message excluding the header.

Figure 10-3
The GTP message
header.

Octets			
1	Version	Spare "1111"	LFN
2	Message Type		
3-4	Length		
5-6	Sequence Number		
7-8	Flow Label		
9	LLC Frame Number		
10	X X X X X X X X X		
11	Spare "11111111"		
12	Spare "11111111"		
13-20	TID Number (MCC, MNC, MSIN + NSAPI)		

■ **Sequence number**   Is a transaction identity for signaling messages and an increasing sequence number for tunneled T-PDUs.

■ **Flow label**   Identifies unambiguously a GTP flow.

■ **LLC frame number**   Is used as the inter-SGSN routing update procedure to coordinate the data transmission on the link between the mobile station and SGSN.

■ **Spare bits x**   Indicates the unused bits, which are set to 0 by the sending side and are not evaluated by the receiving side.

■ *Tunnel identifier* **(TID)**   Points out MM and *Packet Data Protocol* (PDP) contexts. This identifier is composed of the *Mobile Country Code* (MCC), *Mobile Network Code* (MNC), and *Mobile Station Identification Number* (MSIN). These are parts of the *International Mobile Subscriber Identity* (IMSI) and the *Network layer Service Access Point Identifier* (NSAPI), which is an integer value in the range 0 to 15, identifying a PDP context belonging to a specific MM context ID.

GPRS Tunneling Protocol (GTP) Layer

The SGSN interfaces with the GSM network, whereas the GGSN interfaces with the external world. The GTP is used to transfer information between the SGSN and the GGSN. In essence, IP is the protocol between the SGSN and the GGSN.

As mentioned earlier, tunneling refers to the encapsulation of a user's data packet within another packet. The packets that reach the SGSN or the GGSN (maybe of different formats such as IP, X.25, and so on) are encapsulated packets with the source and destination support node addresses in the outer packets' header. As a result, the actual information from the user is not modified. This is useful because it supports multiprotocol packets to be tunneled through the GPRS backbone.

The tunnels are established when an SGSN activates a PDP context with the GGSN. TIDs identify the tunnels. These are shown in Figure 10-4. Every tunnel has a unique TID. SGSN and GGSN tables are mapped according to the TIDs. The tunnel is destroyed when the context is deactivated. Tunneling is supported for inter-PLMN and intra-PLMN communications.

GTP Identities

A many-to-many relationship exists between the SGSNs and the GGSNs. Therefore, multiple tunnels can exist between the SGSN and the GGSN. A TID that is unique for that pair of nodes identifies each tunnel.

Different network applications on the same mobile could use different tunnels between the GPRS support nodes. The tables in the SGSN and the

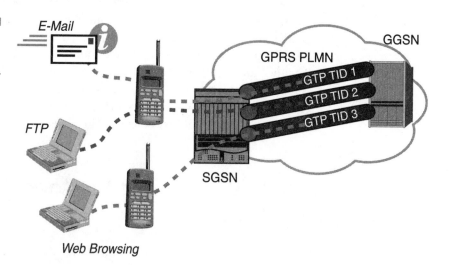

Figure 10-4
The GTP tunnel identities.

GGSN have identifiers that map a particular mobile address with its NSAPI, *Temporary Logical Link Identity* (TLLI), and PDP contexts.

During handover, when a mobile attaches itself to a different SGSN, the queued packets are tunneled to the new SGSN from the old SGSN.

Virtual Private Networks (VPNs)

GPRS must support access to private corporate networks. Corporations expect convenient, but secure, access from wireless data networks. Roaming mobile corporate users should have secure, trusted access to the company's data vaults. The term *Wireless Virtual Private Network* (W-VPN) is used to describe such an environment. We will review key tunneling, authentication, and encryption techniques and ways that GPRS can use them to provide secure corporate network access. Figure 10-5 shows the VPN.

A VPN is an extension of an organization's private intranet across a public network (the Internet), creating a secure connection essentially through a tunnel. VPNs securely convey information across the Internet connecting remote users, branch offices, and business partners into the corporate network.

VPNs are owned by the carriers, but are used by corporate customers as if the customers owned them. A VPN is a secure connection that offers the privacy and management controls of a dedicated point-to-point leased line, but actually operates over a shared routed network.

A VPN provides a corporation with many of the benefits of a dedicated network, without the expense of deploying and maintaining equipment and

Figure 10-5
The VPN.

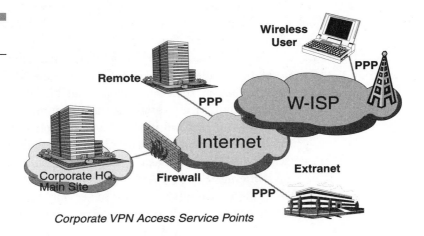

Corporate VPN Access Service Points

facilities. Often VPN solutions include the use of services from telecommunication carriers. The term *Virtual Private Network* originated with voice long-distance deregulation in 1984. In the last 10 years, data VPNs have also become popular and are often implemented using Frame Relay services. Within the past few years, VPN services based on the Internet have become available. With the expansion of IP into our networks, VPNs now span from voice to data services, from wireline to wireless. Operators offer value-added service via connectivity at far lower rates than dedicated leased lines. The GPRS VPN operator provides a range of services from full outsourcing of the corporate data network operation to providing selected parts of it, like remote access, site connectivity, or extranet services with partners. Access by remote mobile workers is becoming more important as telecommuting increases and their productivity gains become an obvious part of the information delivery process. GPRS wireless access services make this possible. Many GPRS operators will enter this lucrative and competitive market. Sometimes the computers we don't see will participate in VPNs: in vehicles, vending machines, and appliances that are remotely monitored where wireline connectivity is not feasible or financially practical. W-VPNs will share resources in the GPRS backbone. GPRS VPNs are based on standard IPs and feature seamless interoperability between providers. Some providers may support multiple access technologies. With intranets everywhere, IP will be the main corporate network protocol, enhanced to provide end-to-end security, confidentiality, authentication, and integrity over a shared VPN network.

The *Password Authentication Procedure* (PAP) and the *Challenge Handshake Authentication Protocol* (CHAP) do little for security. In fact, PAP and CHAP are part of the basic *Point-to-Point Protocol* (PPP) suite and fall short in providing a true security procedure. These schemes do not address issues of ironclad authentication and integrity or eavesdropping. The PAP and CHAP are rudimentary procedures used to log onto a network, but hackers and crackers can easily defeat both.

Layer 2 Tunnel Protocol (L2TP) is another variation of an IP encapsulation protocol. Encapsulating an L2TP frame inside a UDP packet creates an L2TP tunnel. This, in turn, is encapsulated inside an IP packet whose source and destination addresses define the tunnel's ends. Because the outer encapsulating protocol is IP, IPSec protocols can be applied to this composite IP packet, thus protecting the data that flows within the L2TP tunnel. The *Authentication Header* (AH), *Encapsulated Security Payload* (ESP), and *Internet Security Association and Key Management Protocol* (ISAKMP) protocols can all be applied in a straightforward way. L2TPs are an excellent way of providing cost-effective remote access, multiprotocol transport, and remote *Local Area Network* (LAN) access. It does not provide cryptographic robust security. Therefore, L2TP should be used in conjunction with IPSec for providing secure remote access. L2TP supports both host-created and ISP-created tunnels.

A remote host that implements L2TP should use IPSec to protect any protocol that can be carried within a PPP packet. IPSec offers a variety of advantages. The chief among those are the following:

- IPSec is widely supported by the industry including Cisco, Microsoft, Nortel Networks, and so on.

- This universal presence ensures interoperability and availability of secure solutions for different types and kinds of end users. In addition, all IPSec-compliant products from different vendors are required to be compatible.

- IPSec provides for transparent security, irrespective of the applications used.

- IPSec is not limited to operating system-specific solutions, for example. It will be ubiquitous with IP. It will also be a mandatory part of the forthcoming IPv6 standard.

- IPSec offers a variety of strong encryption standards. The key design decision to support an open architecture provides easy adaptability of newer, stronger cryptographic algorithms.

- IPSec includes a secure key management solution with digital certificate support. IPSec guarantees the ease of management and use. This reduces deployment costs in large-scale corporate networks.

- IPSec used in conjunction with L2TP provides secure remote access client-to-server communication. L2TP alone cannot provide for a totally secure communication channel due to its failure to provide per-packet integrity, its incapability to encrypt the user datagram, and the limited security coverage only at the ends of the established tunnel.

The major drawback to packet-filtering techniques is that they require access to clear text, both in packet headers and in the packet payloads. When encryption is applied, some or all of the information needed by the packet filters may no longer be available. For example, in transport mode, ESP will encrypt the payload of the IP datagram. In tunnel mode, ESP will encrypt the entire original datagram, both header and payload.

In most IPSec-based VPNs, packet filtering will no longer be the principle method for enforcing access control. IPSec's AH protocol, which is cryptographic robust, fills that role, thereby reducing the role of packet filtering for further refining after IPSec has encrypted the packet. Moreover, because IPSec's authentication and encryption protocols can be applied simultaneously to a given packet, strong access control can be enforced even when the data itself is encrypted.

Authentication

IPSec has two major drafts: the AH and ESP. They are defined as follows:

- **AH** Is used to provide connectionless integrity and data origin authentication for an entire IP datagram (hereafter referred to as *authentication*).

- **ESP** Provides authentication and encryption for IP datagrams with the encryption algorithm used determined by the user. In ESP authentication, the actual message digest is now inserted at the end of the packet (whereas in AH the digest is inside the authentication).

AH provides data integrity only and ESP, formerly encryption only, now provides both encryption and data integrity. The difference between AH data integrity and ESP data integrity is the scope of the data being authenticated.

AH authenticates the entire packet, whereas ESP doesn't authenticate the outer IP header. In ESP authentication, the actual message digest is now inserted at the end of the packet, whereas in AH, the digest is inside the AH.

The IPSec standard dictates that prior to any data transfer occurring, a *Security Association* (SA) must be negotiated between the two VPN nodes (gateways or clients). The SA contains all the information required for execution of various network security services such as the IP-layer services (header authentication and payload encapsulation), transport- or application-layer services, and self-protection of negotiation traffic.

These formats provide a consistent framework for transferring key and authentication data that is independent of the key generation technique, encryption algorithm, and authentication mechanism.

One of the major benefits of the IPSec efforts is that the standardized packet structure and SA within the IPSec standard will facilitate third-party VPN solutions that interoperate at the data transmission level. However, it does not provide an automatic mechanism to exchange the encryption and data authentication keys needed to establish the encrypted session, which introduces the second major benefit of the IPSec standard: key management infrastructure or *Public Key Infrastructure* (PKI).

The IPSec working group is in the development and adoption stages of a standardized key management mechanism that enables safe and secure negotiation, distribution, and storage of encryption and authentication keys. A standardized packet structure and key management mechanism will facilitate fully interoperable third-party VPN solutions.

Other VPN technologies that are being proposed or implemented as alternatives to the IPSec standard are not true IPSec standards at all. Instead, they are encapsulation protocols that tunnel higher-level protocols into link-layer protocols.

Security

The key technologies that comprise the security component of a VPN are

- Access control to guarantee the security of network connections
- Encryption to protect the privacy of data
- Authentication to verify the user's identity as well as the integrity of the data

Network security is an extremely important issue for network administrators and one that must be addressed in remote access services.

Integrated at the VPN point of access, user authentication establishes the identity of the person using the VPN node. This is because an encrypted session is established between the two locations. The user authentication mechanism gives the authorized user of the VPN system access to the system, while preventing the attacker from accessing the system. Some of the common user authentication schemes are

- Operating system username/password
- S/Key (one-time) password
- *Remote Access Dial-In User Server* (RADIUS) authentication scheme
- Strong two-factor, token-based scheme

The strongest user authentication schemes available on the market are two-factor authentication schemes. These require two elements to verify a user's identity: a physical element in his or her possession (a hardware electronic token) and a code that is memorized (a PIN number). Some cutting-edge solutions are beginning to use biometric mechanisms such as fingerprints, voiceprints, and retinal scans. However, these are still relatively unproven.

When evaluating VPN solutions, it is important to consider a solution that has both data authentication and user authentication mechanisms. Currently, many VPN solutions provide only one form of authentication.

Because of this, VPN solution providers that only support one of the two authentication mechanisms typically refer to authentication generically, without qualification of whether they support data authentication, user authentication, or both. A complete VPN solution supports both data authentication (also known as the digital signature process or data integrity) as well as user authentication (the process of verifying VPN user identity).

Various cryptographic techniques can be used to ensure the data privacy of information transmitted over an unsecured channel such as the Internet, as in the case of a VPN. The transmission mode used in the VPN solution determines which pieces of the message are encrypted. Some solutions encrypt the entire message (IP header and data), whereas others encrypt only the data.

The four transmission modes used in VPN solutions are

- **In-place transmission mode**   This is typically a vendor-specific solution where only the data is encrypted. Packet size is not affected, which ensures that downstream transport mechanisms are not affected.

- **Transport mode**   Only the data is encrypted and the packet size increases in size. This mode provides adequate data privacy for node-to-node VPNs.

- **Encrypted tunnel mode**   The IP header information and the data are encrypted with a new IP address created and mapped to the VPN endpoints, providing excellent overall data privacy.

- **Nonencrypted tunnel mode**   Nothing is encrypted, which means that all data transported is clear text. This is not an effective solution for data privacy.

Oddly enough, some VPN solutions do not perform any encryption at all. Instead, they rely on data encapsulation techniques such as a tunneling or forwarding protocol for data privacy.

Not all of the tunneling and forwarding protocols use a cryptographic system for data privacy. This means that the protocol would transmit all data in the clear, leaving one to wonder how nonencryption-based solutions can provide any form of data privacy protection—a critical requirement for a VPN. Unfortunately, the industry terminology itself may be contributing to some of this confusion. To clarify this issue, one must look specifically at what transmission mode is being used.

As with the qualification between data authentication and user authentication, transmission modes should be distinguished between encrypted and nonencrypted. If a VPN solution does not provide any form of encryption for data privacy, then this solution is more appropriately called a *Virtual Network* (VN) because nothing is private about the network.

Roaming and Wireless VPNs

Wireline roaming services and organizations currently provide worldwide gateway-to-gateway voice over IP services, worldwide local-rate corporate dial-up, and PPP-based network access via their clearinghouse services, enabling dynamic VPN setup across a number of providers. Figure 10-6 is a representation of roaming and W-VPN services. Likewise, it will be natural for wireless operators to extend their roaming agreements to include VPN and corporate network access service over the GPRS network. In the future, global roaming and WWW infrastructure could be deployed, together with a widespread availability of mobile IP services. The same infrastructure could eventually be used for both wireline and wireless services.

Figure 10-6
Roaming and
Wireless VPNs.

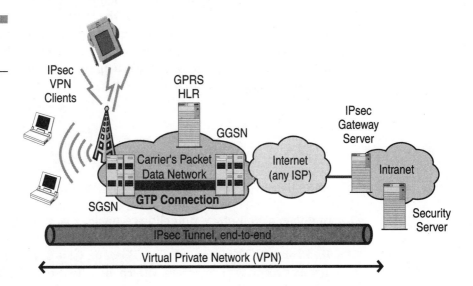

What Makes GPRS VPNs Different?

A GPRS VPN shares many requirements with other VPNs. The remote user needs network access comparable to that of on-premise corporate computers. The remote user must be authenticated, possibly by both the access network and by the corporation. No eavesdropping should occur on data flowing between the remote user and the corporation, nor should it be possible for the data to be altered by a third party. The presence of VPN users and the infrastructure to support them should not provide a conduit for an intruder to breach the corporate firewall. When a GPRS VPN is being considered, a corporation should evaluate several factors unique to the wireless world. Security aspects may be the foremost concern, especially for what might appear to be the most vulnerable portion of the mobile-corporation path: the air link. *Time Division Multiple Access* (TDMA) and the other GPRS access technologies have designed the mobile network so that the packet data traffic is protected by encryption over the air. This improves performance, as many end-to-end encryption methods add extra bytes to each packet sent over the air. They also interfere with the data compression techniques implemented between the mobile system and the GPRS operator's network.

The availability of VPN service for roaming users should be discussed. Some corporations will only want their networks accessed using selected wireless operators or from selected geographical locations. Multinational corporations may decide that roaming users should connect with local VPN access points. The performance of the air link, especially the throughput seen by data users, varies. Although the GPRS air link has multiple methods to ensure reliable air-link data transmission, factors such as fading and multipath may reduce performance. Enhancements to the air link and network infrastructure to meet enhanced *quality of service* (QoS) requirements are underway in standards organizations.

VPN—Service Provider Independent (SPI)

End-to-end, voluntary tunneling technology is used to support corporate-based VPNs. The best example of this is IPSec, an *Internet Engineering Task Force* (IETF) standard. The IPSec tunnel is from a gateway server that matches security parameters with client software on the user's PC, which is distributed by the company to its mobile workers. The IPSec technology provides a secure tunnel extended from the remote clients via the GPRS backbone, across the Internet, and to the gateway via basic TCP/IP protocols. If authentication is successful, the user enters the corporate intranet to access host computers and servers. Voluntary IPSec tunnels include encryption of all data. This places an added overhead of often more than 25 percent on the *route information field* (RIF) interface. Voluntary tunneling is end-to-end and terminates at the user's PC or *personal digital assistant* (PDA), as seen in Figure 10-7. It can be delivered independently of the infrastructure provider. More bandwidth is used over the air interface rather than with compulsory tunneling (see the following list). Other voluntary tunneling examples that are used with the GPRS architecture include

- **Point-to-Point Tunneling Protocol (PPTP)**   From the mobile station to a gateway within the corporate network.

- **L2TP**   From the mobile station to a corporate *L2TP Network Server* (LNS) within the corporation. The mobile station itself takes on the *L2TP Access Concentrator* (LAC) function. Note that L2TP does not itself include data encryption. Running L2TP on top of an IPSec layer provides the needed confidentiality.

These other two voluntary tunneling solutions use PPP. If a compression option for PPP can be negotiated between the endpoints, this provides some compensation for the additional headers required by tunneling.

Service Provider Independent (SPI) voluntary tunneling is a solution technically independent of a service provider's infrastructure, involving

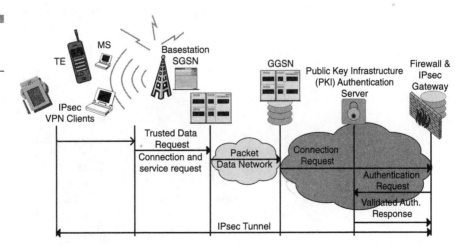

Figure 10-7
SPI voluntary tunneling.

only transport of information. All security aspects are an end-to-end responsibility, not a network responsibility. Nonetheless, wireless carriers will be asked to offer technical support to corporate clients who can build and manage their own end-to-end VPN solutions or outsource to third parties. In a robust SPI implementation, the corporation's tasks would include

- Distribute client software to remote corporate citizens needing secure wireless access
- Manage internal security systems: AAA server, real-time intrusion detection system
- Manage multiple *service level agreements* (SLAs) with service providers

VPN—IPSec End-to-End with PKI

Voluntary-based W-VPN access depicts the session flow of IPSec in an end-to-end service context, as shown in Figure 10-8. Use of a trusted third-party *Certificate Authority* (CA) with its PKI is exposed. This solution might utilize a manufacturer's access router or managed firewall products (Lucent, Cisco, Nortel, and so on) based on IPSec techniques and compatible with the industry's RADIUS and PKI products.

The GPRS operator may provide portions of the PKI. Although its role in providing the VPN might seem limited, the operator can play an important role to reduce security risks associated with the SPI voluntary tunnels. Although the remote user can have a secure path to the corporate intranet, the user's device may still be accessible from the larger, insecure Internet. If that device were compromised, the security of the corporation would be

Figure 10-8
VPN—IPSec end-to-end.

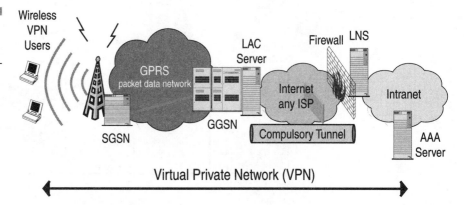

imperiled. The problem stems from having a remote device that has a network presence on both secure and insecure networks. The addition of protective firewall software on the client is one potential solution. Another solution would be for the GPRS operator to screen the mobile's traffic to/from IP addresses other than the corporate gateway.

VPN—Service Provider Dependent (SPD)

GPRS W-VPN can use a compulsory tunneling technique. Rather than an end-to-end solution, the VPN is made of two separate pieces. One is the GPRS operator's access network and another traverses the Internet. The network operator delivers a value-added, secure, access service. Its access concentrator is the tunnel establishment point. After connecting to it, a user's data is tunneled to a termination device at the edge of the corporate network. At this point, user packets are authenticated and access to corporate data is authorized, as shown in the flow in Figure 10-9. The carrier builds and manages this network-to-network compulsory solution. Example compulsory GPRS solutions include

- LAC in a carrier's network between GGSN and the corporate firewall
- Gateway-to-gateway compulsory IPSec tunneling
- A dedicated GPRS-GGSN, colocated at/near the corporate firewall (using GTP tunneling)

L2TP-Based Wireless VPN in a GPRS Infrastructure

Standardized L2TP (RFC2661) evolved from various proprietary protocols and is intended to improve interoperability between different vendors' tunneling equipment. L2TP is not an end-all answer. Notably, L2TP does not provide its own security (encryption), but can make use of IPSec. L2TP operates in compulsory mode in which the tunnel is supported by service provider gear. L2TP can also operate in a voluntary mode, in which the tunnel endpoint is actually on the user's laptop. In this case, the user incurs the tunnel overhead.

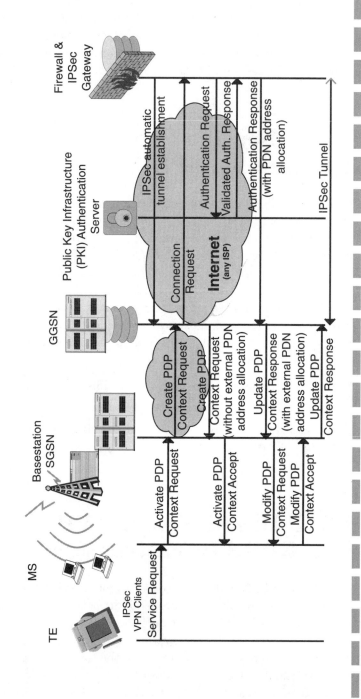

Figure 10-9 *VPN—SPD.*

The mobile device requests a PPP session that starts at the mobile and then exits the GPRS network at the GGSN. The GGSN either contains or is collocated with an L2TP LAC. The LAC tunnels the PPP stream (via L2TP packets carrying PPP frames) to an L2TP LNS at the corporate network's edge. Actual authentication and authorization, as well as IP address assignment, occurs here, under the control of the corporate network.

This scenario is based on an SLA or SA established between an LAC function that is at the GGSN and corporate-based LNS. The SA provides privacy rules and L2TP tunnels for user data. This is important because L2TP does not support its own security and because PPP encryption is neither widely deployed nor has multivendor interoperability. In this compulsory scenario, the operator assigns a corporation an *Access Point Name* (APN) Network Identifier. The APN is used by the SGSN to select the GGSN to be addressed for a specific group of corporate mobile users.

The GGSN receives the IP address of the LNS at the user's corporate network. This is passed to the LAC function for packet forwarding to the LNS. The user accesses a corporate network after his or her wireless device first attaches to the GPRS network, using a data type PPP and specifying the APN. Once the PDP context is active, control is passed to the LAC so it can relay the information used by PPP. This triggers the establishment of an L2TP control connection to the corporate LNS. If an L2TP tunnel is already established for the corporate VPN connection, the newly attached user can share it. If not, a new tunnel is created. The GGSN then uses the L2TP control connection to establish an L2TP call (L2TP tunnel to carry PPP) between the LAC and the LNS. The authentication of the mobile is performed via the corporate LNS. The corporate LNS often utilizes the services of the corporate AAA system (for example, RADIUS). After the authentication phase, an IP address is assigned to the mobile, as is normally the case. The mobile does not have a carrier-network IP address associated with it; instead, the mobile has established a PPP session directly with the corporate network.

IPSec Gateways and Compulsory Tunneling W-VPN

This example is similar to the previous example in the L2TP section, except the corporation's VPN gateway server establishes an IPSec rather than an L2TP tunnel. In the previous section, the SA between the LAC and LNS was pre-provisioned. In the IPSec case, shown in Figure 10-10, the SA could be dynamically established, but it is more likely to be pre-provisioned because administration is more complex. The tunnel goes between the two gateways and is not an end-to-end solution.

It is important to note that this solution requires that the GGSN directly associates the mobile's IP address with the IPSec tunnel so that packets to/from the mobile traverse only that tunnel. The corporation may internally use IP addresses that conflict with normal Internet assignments through the use of private IP addresses as described in RFC1918, for example.

This scenario may be viable for corporations seeking to lower the cost of their internal security systems, while avoiding full trust in their wireless carrier. Or, the wireless carrier can offer to provide this PKI as a value-added service, perhaps including wireless e-commerce transaction services for horizontal services/markets.

Multiple VPN Gateway Architecture

Multiple VPN gateway architectures may also be used in a GPRS backbone system. In this example, each VPN can be defined by filtering rules in a managed firewall, as shown in Figure 10-11. The previous example suggests a trusted GPRS *Home Location Register* (HLR) lookup as the means

Figure 10-10
Compulsory tunneling with VPN.

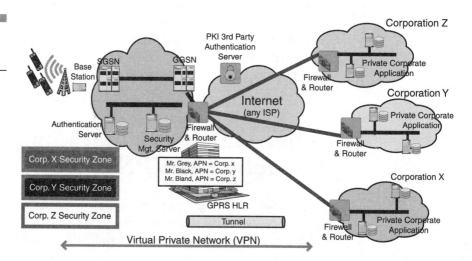

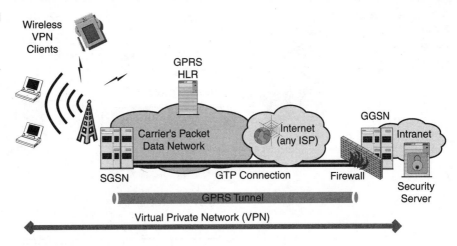

Figure 10-11
Multiple VPN
gateway architecture.

to authenticate a mobile subscriber to a VPN group and be given access behind its corporate firewall.

Putting a GGSN at or behind the corporate firewall provides GTP tunneling over the Internet. With the addition of an IPsec point-to-point tunnel between the GPRS network and the GGSN, the payload itself is protected. The remote GGSN case is an interesting alternative with a router-based platform for the GGSN. Certain issues still need to be addressed including

- Is it cost-effective to have a GGSN on the corporate premises?
- How does the service provider ensure the security of the GPRS infrastructure when a connection to the SGSN, a key network node, exists in a remote corporate site?
- Who is responsible for ownership and administration of the GGSN?

Using the VPN Tunnel

In the example shown in Figure 10-12, a tunnel is created between the GGSN and the intranet switch. The mobile station sends IP packets just as if it were located in the remote intranet. The source IP packets sent by the mobile station are encapsulated by the GGSN within other IP packets. The new header contains the GGSN IP address as the source address and the IP address of the intranet switch as the destination IP address.

The destination switch at the end of the tunnel performs the opposite operation and the original IP packets are forwarded to the target host. In

Figure 10-12
Using the VPN
tunnel.

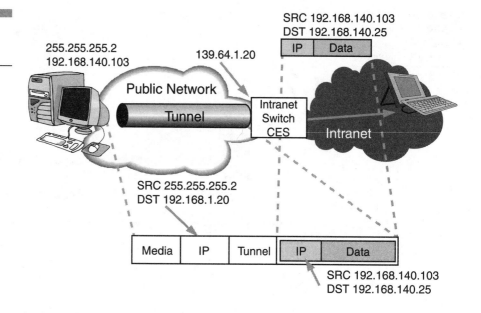

case of an insecure network between the GGSN and the intranet switch (for example, the Internet network), a security protocol like IPSec can be used.

The four existing tunneling protocols are the PPTP, L2F, L2TP, and IPSec:

- PPTP, L2F, and L2TP are L2TP; they are connectivity-oriented protocols that are limited in supporting PPP connections. This makes them unavailable for LAN-to-LAN connections.

- PPTP is a client-initiated protocol that doesn't require special ISP service; it creates a tunnel toward the destination switch using an IP in IP encapsulation and a *generic routing encapsulation* header (GRE).

- L2F and L2TP do not require any software on the client; they are implemented at the ISP side. A tunnel is created from the ISP *Network Access Server* (NAS) or the ISP LAC to the L2F or LNS of the destination switch. L2TP is a synthesis of PPTP and L2F protocols.

- IPSec is a Layer 3 Tunneling Protocol; it is a security-oriented protocol. It provides security protocols, including an AH and an encapsulation security header (for encryption), SAs to provide security for a connection data, key management, encryption, and an authentication algorithm.

PDP Context SGSN Role

The SGSN receives from the mobile station an Activate PDP context with the optional values PDP type, PDP address, and APN. Figure 10-13 shows the information passed along the link.

The SGSN compares the received values with its local subscribed values (this information exists locally because of the GPRS attach). If the PDP context is valid, the SGSN has to perform some operation like the following:

- Create a tunnel TID by combining the IMSI and the NSAPI.
- Send a query to the *Domain Name System* (DNS) server to obtain the IP address of the GGSN using the complete APN (with the Operator Identifier). The DNS server answers by giving it a list of IP addresses.
- Send a Create PDP Context Request message to the GGSN with the PDP type, PDP address, APN Network Identifier, TID, selection mode, and so on.

The APN is made of two parts:

- APN Network Identifier
- APN Operator Identifier

The structure is MNCyyy.MCCzzz.gprs (depending on the IMSI).

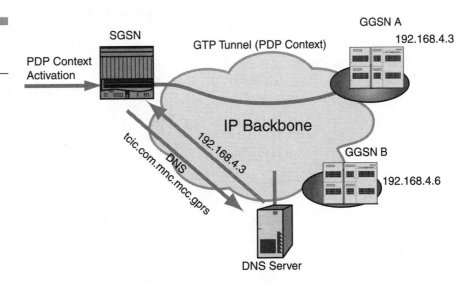

Figure 10-13
Using the PDP context SGSN role.

Create PDP Context GGSN Role

When receiving the Create PDP Context Request message coming from the SGSN, the GGSN uses the selection mode and the APN Network Identifier to validate the PDP Context. Figure 10-14 shows the role of the GGSN. According to the network configuration mode, the GGSN has to perform the following:

- DNS query to find the IP address of the remote switch
- Mobile IP @ assignment from the local *Dynamic Host Configuration Protocol* (DHCP) server or the remote DHCP or RADIUS server
- Tunnel creation toward the remote switch with a security protocol in case of public external network
- Authentication and authorization checks

When these operations are complete, the GGSN returns a Create PDP Context Response message to the SGSN with the assigned IP address, if applicable.

Information describing the various users, applications, files, printers, and other resources accessible from a network is often collected into a special database that is sometimes called a directory. As the number of different networks and applications has grown, the number of specialized directories of information has also grown, resulting in islands of information that are difficult to share and manage. If all of this information could be maintained and accessed in a consistent and controlled manner, it would

Figure 10-14
The GGSN role.

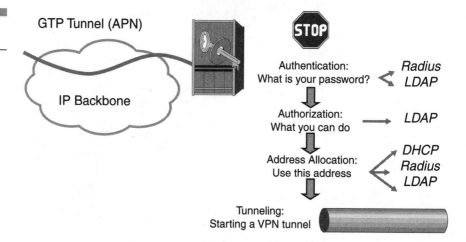

provide a focal point for integrating a distributed environment into a consistent and seamless system.

The *Lightweight Directory Access Protocol* (LDAP) is an open industry standard that has evolved to meet these needs. LDAP defines a standard method for accessing and updating information in a directory. LDAP gained wide acceptance as the directory access method of the Internet and is therefore strategic within corporate intranets. It is supported by a growing number of software vendors and is being incorporated into a growing number of applications. For example, the two most popular Web browsers, Netscape Navigator/Communicator and Microsoft Internet Explorer, support LDAP functionality as a base feature.

What Is a Directory?

A directory is a collection of information about objects arranged in some order that gives details about each object. Popular examples are a city telephone directory and a library card catalog. For a telephone directory, the objects listed are people; the names are arranged alphabetically and the details given about each person typically include an address and telephone number. Books in a bookstore are ordered by author or by title, and information such as the ISBN number of the book and other publication information is also contained.

In computer terms, a directory is a specialized database (also called a data repository) that stores typed and ordered information about objects. A particular directory might list information about printers (the objects) consisting of typed information, such as location, speed in pages per minute (numeric), print data streams supported (for example, PostScript or ASCII), and so on. Directories enable users or applications to find resources that have the characteristics needed for a particular task. The LDAP is used with the corporate intranet so that the tunneling wireless users can access the various services on the intranet.

The words "local," "global," "centralized," and "distributed" are often used to describe a directory or directory service. These words mean different things to different people in different contexts. In this section, these terms are explained as they apply to directories in different contexts. In general, *local* means something is close by, and *global* means that something is spread across the universe of interest. The universe of interest might be a company, a country, or the carrier's PLMN. Local and global are two ends of a continuum. That is, something may be more or less global or local than something else. Like local and global, something can be distributed to a greater or lesser extent.

The information stored in a directory can be local or global in scope. For example, a directory that stores local information might consist of the names, e-mail addresses, and public encryption keys of members of a department or workgroup. A directory that stores global information might store information for an entire company. In this context, the universe of interest is the company. The clients who access information in the directory can be local or global. Local clients may all be located in the same building or on the same LAN. Global clients might be distributed across the continent or PLMN.

Transparent Access

Transparent access (shown in Figure 10-15) means that the GGSN doesn't participate in user authentication; the GPRS authenticates based on GSM authentication of the user's *Subscriber Identity Module* (SIM) card only. The GGSN is able to allocate the user a public IP address; this user has to subscribe to the ISP for Internet services such as e-mail, news, Web, and so on.

In most cases, the transparent mode is used when the GPRS operator already is an ISP. In this case, the GPRS operator provides basic Internet

Figure 10-15
Transparent access.

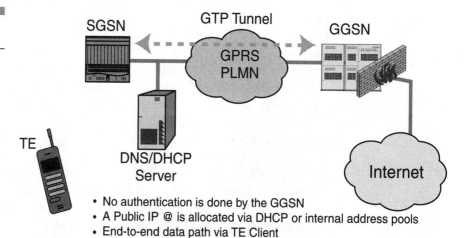

- No authentication is done by the GGSN
- A Public IP @ is allocated via DHCP or internal address pools
- End-to-end data path via TE Client
  - IPsec for data security
  - PPP, PPTP, L2TP for user authentication

access. For this reason, the GPRS operator allocates a static or dynamic public address to the mobile station using the DHCP local server.

Transparent Mode

In case of intranet access, the authentication of this user will be performed at the service provider side. Figure 10-16 shows this in the transparent mode.

A tunnel is created from the user to the remote switch of the destination network. This tunnel may be encrypted using IPSec, if the communication between the GPRS PLMN and the destination network is performed over an insecure network.

As the security is ensured on an end-to-end basis between the mobile station and the intranet by the *Intranet Protocol*, no specific security protocol exists between GGSN and the intranet. User authentication and the encryption of user data are done within the Intranet Protocol and this protocol can also carry out private IP addresses allocated by the intranet. The basic principles of the transparent mode are

- The GPRS operator allocates a static or dynamic public address to the mobile station static using the DHCP local server.

- The ISP/intranet using RADIUS performs the authentication and authorization.

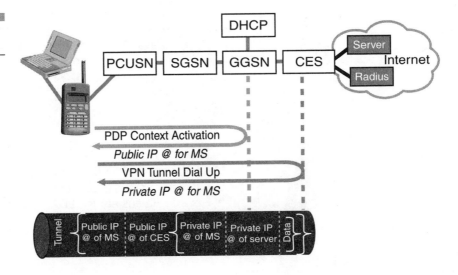

Figure 10-16
Transparent mode.

- The Intranet Protocol running between the user and the ISP/intranet provides end-to-end security.

- A second (internal) IP address may be assigned to the user by the destination network.

Nontransparent Access

In the following example, the GGSN provides an interconnection to the intranet, as shown in Figure 10-17. In nontransparent GPRS access, the GGSN facilitates the user access to the ISP or the intranet. The mobile station is allocated a private static or dynamic IP address belonging to the ISP or intranet; the GGSN has to send a query to the ISP/intranet RADIUS/DHCP server. The authentication and authorization are performed by the GGSN on the RADIUS server belonging to the destination network. A tunnel is created from the GGSN to the ISP or intranet switch.

Nontransparent Mode

As the interconnection between the GPRS network and the destination network may be insecure, a security protocol may be defined by mutual

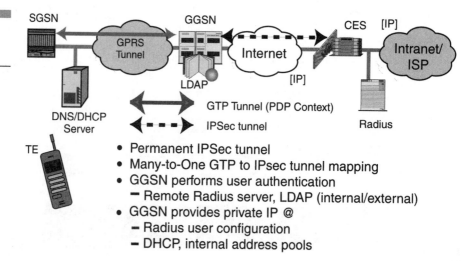

Figure 10-17
Nontransparent access.

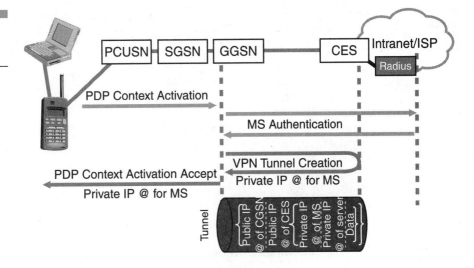

Figure 10-18
Nontransparent mode.

agreement between the two parts. Figure 10-18 shows the basis of non-transparent mode.

Receiving the PDP Context Request message from the SGSN, the GGSN deduces the following from the APN:

- The server to be used for address allocation and authentication
- The protocol to be used with those servers, such as RADIUS, DHCP, and so on
- The communication and security feature needed to dialogue with those servers, such as tunnel type or IPSec SA

When the GGSN sends the Create PDP Context Response message to the SGSN, the tunnel toward the remote switch is created.

Virtual Dial-Up (Enhanced Nontransparent)

In the virtual dial-up (enhanced nontransparent) access mode (shown in Figure 10-19), the different functions (tunneling, user authentication, security encryption) are provided by different protocols:

- The L2TP session performs tunneling functions.
- Dynamically allocated L2TP sessions take place.

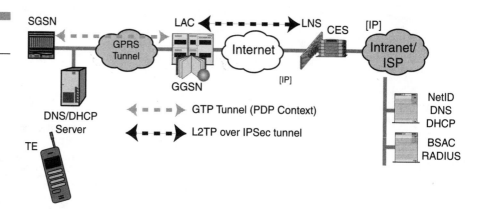

Figure 10-19
Virtual dial-up.

- A permanent IPSec SA is used.
- A one-to-one GTP tunnel to L2TP session mapping is used.
- GGSN is linked to *Circuit Emulating Switch* (CES) via PPP/L2TP.
  - GGSN is LAC.
  - CES is LNS.
- LNS provides user configuration.
  - Authentication (RADIUS, LDAP, proprietary).
  - Private IP address (remote DHCP RADIUS).
- The PPP connection performs user authentication.
- IPSec SA performs IP data frame encryption.

As seen in this chapter, the application is what really makes GPRS work. Although some of the scenarios played out in this chapter are not yet ready to roll out, the concept is exactly what corporate users and providers will negotiate in the future. It is through the converged application and technology that will make or break the acceptance of GPRS in the beginning stages of implementation.

Future Enhancements and Services

Objectives

Upon completion of this chapter, you should be able to

- Describe the concept of the HSCSD network.
- Discuss the evolution of EGPRS and EDGE.
- Understand what UMTS is all about.
- Describe the various other enhancements of 3G.
- Understand how ETSI and ANSI are collaborating with the developments.

Mobile Evolution

Get ready! As the convergence of wireless technology and the Internet continue at an escalating pace, the new possibilities created by *third-generation* (3G) and *fourth-generation* (4G) technologies appear endless. Preparing for the revolution, existing *Time Division Multiple Access* (TDMA) operators must evolve their networks to take advantage of mobile multimedia applications and the eventual shift to an all-*Internet Protocol* (IP) architecture. One way to do that is through the evolution of *General Packet Radio Service* (GPRS) and *Enhanced GPRS* (EGPRS). However, soon after we see the installation of GPRS, some operators will begin the next step in the evolution process to *Enhanced Data* rates for *Global Envolution* (EDGE). With EDGE, existing TDMA networks can host a variety of new applications, including

- Online e-mail
- Access to the World Wide Web
- Enhanced *Short Message Services* (SMSs)
- Wireless imaging with instant photos or graphics
- Video services
- Document/information sharing
- Surveillance
- Voice messaging via Internet
- Broadcasting

At the same time, some operators will skip the step to EDGE and go directly to *Universal Mobile Telecommunications Systems* (UMTSs), or what we consider to be a 3G technology. Figure 11-1 shows the steps, as the carriers choose which way to proceed.

Using a timeline, the evolution of wireless to 3G systems is shown in Figure 11-2, showing the evolution of the various techniques that emerged over the years.

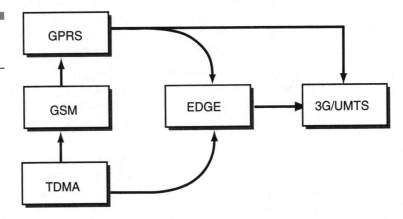

Figure 11-1
The evolution to UMTS choices.

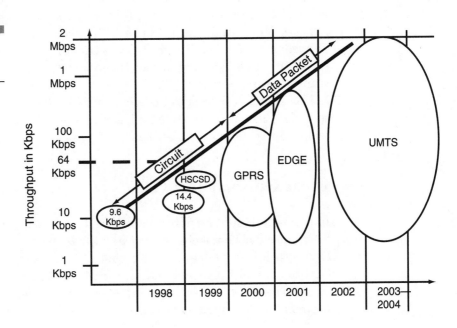

Figure 11-2
Timeline for 3G/UMTS.

Why is packet data technology important? Packet networks provide a seamless and immediate connection to the Internet or corporate intranet, enabling access to existing Internet applications, such as e-mail and Web browsing, without dialing into an *Internet Service Provider* (ISP). The advantage of a packet-based approach is that GPRS uses the medium (in this case, the radio link) only as long as data is being sent or received. Multiple users can share the same radio channel very efficiently. In contrast, with current circuit-switched connections, users have dedicated connections during their entire call, even if they are not sending data. Many applications have idle periods during a session. With packet data, users will only pay for the amount of data they actually communicate, and not the idle time. In fact, with GPRS, users could be "virtually" connected for hours at a time and only incur modest connect charges.

Although packet-based communications works well with all types of communications, it is especially well suited for frequent transmission of small amounts of data. We refer to this as short and bursty, such as real-time e-mail and dispatch (vehicles and field service). Packet is equally well suited for large batch operations and other applications involving large file transfers. However, when using large file transfers, the cost can become very expensive compared to circuit-switched data transmissions. GPRS supports the IP as well as the X.25 protocol. IP support is increasingly more important as companies look to the Internet as a way for their remote workers to access corporate intranets. This is true when using a *Virtual Private Network* (VPN). In the case of VPNs, GPRS works well because of its *GPRS Tunneling Protocol* (GTP), which can secure the mobile data while in transit on the wireless networks, and IPSec transfers can be used when transiting the wireline networks.

UMTS is a *European Telecommunications Standards Institute* (ETSI) term for a 3G mobile telecommunication service. Over recent years, mobile telephony evolutions have become known as the following categories.

First Generation (1G)

In the early 1980s, the *first-generation* (1G) technologies were the world's first public mobile telephone services such as the *Advanced Mobile Phone Service* (AMPS) (United States), *Total Access Communication Services* (TACS) (United Kingdom), and *Nordic Mobile Telephone* (NMT) (Scandinavia). These systems were analogue, provided national coverage (though far from complete in most cases), and offered limited services.

Second Generation (2G)

GSM is by far the world's primary *second-generation* (2G) system. Designed by a joint effort from manufacturers, regulators, and service suppliers from many (European) countries, GSM became a European and then a global standard. *Code Division Multiple Access* (CDMA) systems now under the collective term of *cdmaOne* are the other major 2G technology. Globally, arguments about which technology was superior became largely academic because GSM was deployed first (early 1990s) and rapidly gained universal acceptance (with the exception of the United States and Japan). CDMA was launched more recently (mid-1990s) and has shown remarkable uptake and growth. In late 1998, CDMA had an estimated 12 million users and GSM had over 100 million users. Of course, in 2001, CDMA had 95 million CDMA users and GSM had 550 million users!

2G systems offer

- Open standards (arguable for CDMA)
- Digital technology
- Near national coverage and roaming
- Voice and data (limited rates)
- Supplementary services

Third Generation (3G)

The world's leading telecommunication authorities such as *the International Telecommunications Union* (ITU), ETSI, and others are formulating specifications for the next generation of mobile telecommunication devices and networks, as seen in Figure 11-3, as the evolution to mobile 3G. Within ETSI, this network is known as the UMTS and is data focused.

HSCSD

It is intended that *High-Speed Circuit-Switched Data* (HSCSD) will use the 14.4-Kbps channel coding option and that it will additionally use multiple time slots. To see how this might operate requires a basic knowledge of the physical structure of a traffic channel on the air interface.

Figure 11-3
Evolution of mobile
3G by international
standards.

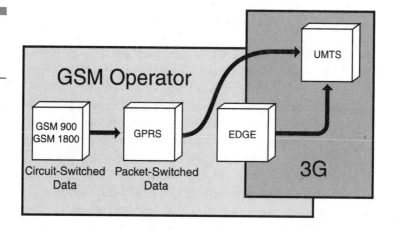

The uplink and downlink of a GSM traffic channel take place on differ-ent frequencies. Also, the uplink and downlink time slots occur at different times in the eight time slot frame. Additionally, when engaged in a traffic channel, a GSM mobile station must constantly be monitoring downlink power levels from neighboring cells as part of the handover process. Over an eight time-slot frame therefore a mobile will

- Receive a downlink burst.
- Transmit an uplink burst.
- Monitor a downlink transmission from a neighboring cell.

Two Time Slots

One restriction that HSCSD places upon multiple time slot links is that the time slots allocated must be consecutive.

As seen in Figure 11-4, the use of two time slots is relatively simple to implement. The mobile is still able to run through its standard routine of receive, transmit, and monitor a neighbor within an eight time slot frame. With three or more time slots being used, an overlap occurs between the receive and transmit times and implementation of this involves substantial hardware changes in the mobile station, that is, the use of a *radio frequency* (RF) duplexer. (At first sight, it looks as if no overlap occurs when using three time slots, but overlap does occur due to the timing advance applied to the uplink.)

Figure 11-4
The two time slots
used in HSCSD.

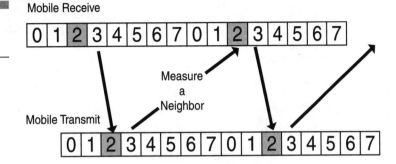

Mobile Receive

Mobile Transmit

Measure a Neighbor

Data Rate—28.8 Kbps

Enhanced General Packet Radio Service (E-GPRS)

It is proposed that EGPRS will offer eight additional coding schemes. The lower layers of the user data plane, which has been specifically designed for GPRS operation, is reflected in the protocol stack, comprising the physical, *Radio Link Control/Medium Access Control* (RLC/MAC), and *Logical Link Control* (LLC) layers. Although the LLC layer can be used without modifications when EDGE functionality is introduced, it is necessary to modify the RLC/MAC layer to support features such as efficient multiplexing and link adaptation. The basic modifications needed for EDGE consider the form of the data blocks that are being transferred across the radio interface. For EGPRS, several combinations of interleaving and coding have been proposed, whereas in the current GPRS proposals, the interleaving depth is set to four bursts.

Link adaptation offers mechanisms for choosing the best modulation and coding alternative for the current radio link. In GPRS, only the coding scheme can be altered between two consecutive LLC frames; however, with EGPRS, a refined link adaptation concept can be utilized that enables both coding and modulation schemes to be changed to suit the given radio link.

In addition, link adaptation should enable seamless switching between the two modulation schemes to such an extent that in EGPRS, the *uplink state flag* (USF) information can be modulated using *Binary Offset Quadrature Amplitude Modulation* (B-O-QAM) and the user data by *Quaternary Offset Quadrature Amplitude Modulation* (Q-O-QAM). B-O-QAM is used in this case for the broadcast purposes and facilitates the characteristics of being robust and therefore available in the entire GSM coverage area.

Table 11-1

The Six Coding
Schemes for EGPRS

Service	Code Rate	Modulation Used	Gross Rate (In Kbps)	Radio Interface Rate (In Kbps)
EGPRS PCS-1	0.326	8-PSK	69.2	22.8
EGPRS PCS-2	0.496	8-PSK	69.2	34.3
EGPRS PCS-3	0.596	8-PSK	69.2	41.25
EGPRS PCS-4	0.756	8-PSK	69.2	51.6
EGPRS PCS-5	0.829	8-PSK	69.2	57.35
EGPRS PCS-6	1.00	8-PSK	69.2	69.2

Six coding schemes have been specified for *eight-phase shift keying* (8-PSK) modulation with regards to EGPRS. These can be seen in Table 11-1. It is assumed that each communication link will be able to choose the modulation and coding combination that achieves the highest throughput for that particular link quality. For example, users with a low *channel-to-interference* (C/I) ratio will be able to perform a link adaptation towards *Gaussian minimum shift keying* (GMSK) as opposed to 8-PSK. This link adaptation between GMSK and 8-PSK should be seamless because both modulation schemes utilize the same symbol rate of 270.833 *kilo symbols per second* (Ksps).

Enhanced Data Rates for GSM Evolution (EDGE)

Beyond GPRS, EDGE takes the cellular community one step closer to UMTS. It provides higher data rates than GPRS and introduces a new modulation scheme called 8-PSK. The TDMA community also adopted EDGE for their migration to UMTS. The data rates allocated for EDGE start at 384 Kbps and above as a second stage to GPRS. EDGE uses the same modulation techniques as many of our existing TDMA infrastructures by using GMSK 8-PSK. Moreover, EDGE uses a combination of *Frequency Division Multiple Access* (FDMA) and TDMA as the multiple access control methods. If we look at this from an OSI stack model, EDGE uses FDMA and TDMA

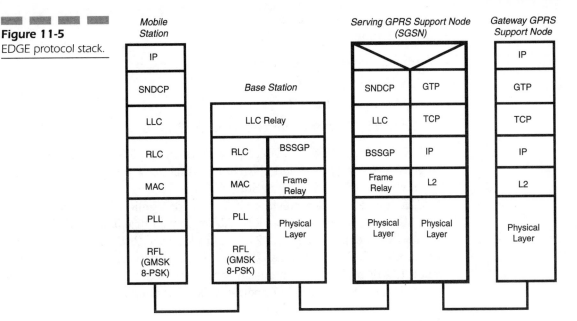

Figure 11-5
EDGE protocol stack.

at the MAC layer (bottom half of Layer 2 in the OSI). Figure 11-5 shows the protocol stack for EDGE.

The channel separations are 45 MHz and the carrier spacing is a 200-kHz channel capacity, the same as GSM and GPRS. The number of TDMA slots on each carrier is the same (eight) as the GSM and GPRS architecture. When a mobile station wants to transmit its data, it can request and use one to eight time slots per TDMA frame. Connectivity is handled via a packet-switched data network such as IP and X.25. These can be public data networks or private data networks.

Although most carriers and service providers have plans to deploy enhanced mobile wireless services at higher speeds, the rollout of high-bandwidth wireless transport technology still faces many possibilities. On a positive note, widespread demand will be sufficient enough to support cellular enhancements like high-speed data services and expanded voice capacity. Competitive pressures will also compel service providers to upgrade. The *International Telecommunication Union-Radiocommunications Standardization Sector* (ITU-R) has actually established five different standards that fall into the category of 3G/UMTS. Moreover, the telecommunications industry is growing increasingly impatient to test the world markets for high-bandwidth wireless communication services. The ITU's *International Mobile Telecommunications-2000* (IMT-2000) initiative may one day converge, but today, many 3G proposals are still under consideration including

- cdma2000 (an upgrade to cdmaOne)
- UMTS
- *Wideband-CDMA* (W-CDMA)
- *Universal Wireless Communications* (UWC-136)

UWC-136 is based on TDMA, just like Europe's GSM, Japan's *personal digital cellular* (PDC) and the *Digital Advanced Mobile Phone System* (D-AMPS) used in the United States.

Existing 2G service providers have already applied for licenses to operate 3G networks around the globe. Although it's unclear what 3G technologies will be adopted, the most 2.5G upgrades are GPRS and HSCSD, which is an upgrade being considered by some GSM network operators. Beyond that, EDGE modulation extensions are planned, which will enable service providers to offer even higher performance, enabling true 3G-like services.

The ITU currently embraces various proposed schemes to attain the IMT-2000 3G vision. From TDMA-based 2G providers of GSM and *North American dual-mode cellular* (NADC) services, interim upgrades will come in the form of GPRS, HSCSD, and IS-136+, and will eventually converge at EDGE for the next throughput upgrade (to 384 Kbps) before 3G.

What Is Special about EDGE?

EDGE is a proposed modification to the modulation scheme utilized by GSM. EDGE is a new modulation scheme that is more bandwidth efficient than the prefiltered GMSK modulation scheme used in the GSM standard. It provides a promising migration strategy for HSCSD and GPRS. The technology defines a new physical layer: 8-PSK modulation, instead of GMSK. 8-PSK enables each pulse to carry 3 bits of information versus the GMSK 1-bit-per-pulse rate. This change will drastically increase the bit rates available to end users for the purpose of data transfer. The expectation is that the enhanced modulation techniques will make it possible to maintain a good quality link by automatically adapting to the radio interference conditions and thereby provide the highest possible rate. The exact implementation and technical details are still being discussed in various ETSI feasibility studies, but one can almost assume that certain factors are near completion. Therefore, EDGE has the potential to increase the data rate of existing GSM systems by a factor of three.

EDGE retains other existing GSM parameters, including a frame length, eight time slots per frame, and a 270.833-kHz symbol rate. The GSM 200-kHz channel spacing is also maintained in EDGE, enabling the use of existing spectrum bands. This fact is likely to encourage deployment of EDGE technology on a global scale. Two additional modulation schemes have been proposed and these are Q-O-QAM and B-O-QAM. These two new modulation schemes will both result in symbol rates of 361.111 Kbps, but Q-O-QAM will offer a higher bit rate as it supports 2 bits per symbol.

Wherever possible, EDGE adopts the GSM standards so as to minimize the changes required by manufactures and operators who want to support this new technology. Figure 11-6 shows a comparison of the two systems. This includes maintaining the same frequency plan, meaning that 200 kHz will still separate carriers. In addition, the TDMA structure supported by GSM will remain intact at eight time slots per frame. Also, the relationship between logical and physical channels will remain unchanged.

The feasibility study carried out by ETSI on EDGE proposes that it will be able to support both transparent and non-transparent circuit-switched services, in addition to the packet-based GPRS. These three new services will be called

- **ECSD-T** *Enhanced Circuit-Switched Data-Transparent*
- **ECSD-NT** *Enhanced Circuit-Switched Data-Non-Transparent*
- **EGPRS** Enhanced GPRS

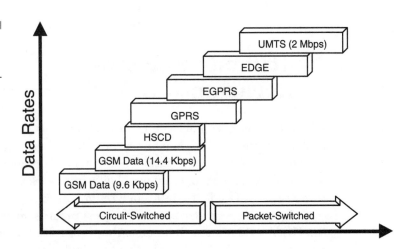

Figure 11-6
Comparing EDGE and GSM.

The Third Generation

The 3G system can be viewed as two distinct elements: the *Access Network* (AN) and the *Core Network* (CN). These two elements have distinct functions and, as far as possible, are independent of each other. This is consistent with the *family of technologies* concept, which enables flexible interconnection between the CN and different access technologies.

UMTS Terrestrial Radio Access Network (UTRAN)

The ETSI decision in January 1998 on the radio access technique for UMTS combined two technologies—W-CDMA for paired spectrum bands and *Time Division Duplex-CDMA* (TD-CDMA) for unpaired bands—into one common standard. This powerful approach promises an optimum solution for all the different operating environments and service needs. The transmission rate capability of UTRA will provide at least 144 Kbps for full-mobility applications in all environments; 384 Kbps for limited-mobility applications in the macro- and microcellular environments; and 2.048 Mbps for low-mobility applications particularly in the micro- and picocell environments. The 2.048-Mbps rate may also be available for short-range or packet applications in the macrocellular environment, depending on deployment strategies, radio network planning, and spectrum availability.

Multimode Second-Generation/UMTS Terminals

UMTS terminals will exist in a world of multiple standards and this will enable operators to offer maximum capacity and coverage to their user base by combining UTRA with second- and other 3G standards. Therefore, operators will need terminals that are able to interwork with legacy infrastructures such as GSM/DCS1800 and *Digital European Cordless Telephony* (DECT) as well as other 2G worldwide standards such as those based on the U.S. AMPS standard, because they will initially have more complete coverage than UMTS. Many UMTS terminals will therefore be multiband

and multimode so that they can work with different standards, old and new. Achieving such terminals at a cost that is comparable with contemporary single mode 2G terminals will become possible because of technological advances in semiconductor integration, radio architectures, and software radio.

This part of the network consists of the *Base Station System* (BSS) and the necessary functionality to control access through the BSS. All radio functionality is contained within the UTRAN. Users' equipment will access the UTRAN with one of the three access modes:

- *Frequency Division Duplex* (FDD) nonsynchronous CDMA
- FDD synchronous multicarrier CDMA
- TDD TD-CDMA

The access mode used will depend on location, conditions, user requirements, and the operator's selected technology options. It is likely that UTRAN will not be the only AN connected to the CN. Subscribers may also have access via fixed, satellite, or cordless connections.

Open Interfaces

If a range of interconnection possibilities for different technologies from different manufacturers is available, then it becomes important to define the various interfaces. Given the general architecture of UMTS, two interfaces will be of particular importance to the ANs. These will be the interface between the *User Equipment* (UE) and the UTRAN, and the interface between the UTRAN and the CN. The interface between the UE and the UTRAN is referred to as the U_u interface and is the radio link using an appropriate multiple access protocol. The interface between the UTRAN and the CN is referred to as the l_u interface. The interfaces in Figure 11-7 can be seen as a simple approach to describing the architecture.

UTRAN Architecture

The UTRAN itself can be subdivided into two different logical elements, as shown in Figure 11-8. These two elements have defined functions and may (or may not) be implemented in different physical elements. These two elements are the *Radio Network Controller* (RNC) and the Node B.

Figure 11-7
The architecture of
3G systems.

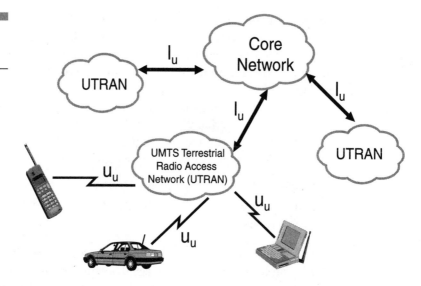

Figure 11-7
The architecture of
3G systems.

Figure 11-8
The UTRAN
architecture.

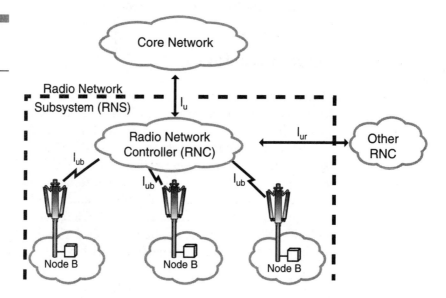

Node B

The Node B is a logical element responsible for radio transmission and
reception in one or more cells. In this respect, it could be physically implemented as a base station.

Radio Network Controller (RNC)

As its name suggests, this is a control element. It is responsible for radio functionality and control of one or more Node Bs. The interface between the RNC and its associated Node B is called the lub interface. The RNC may be physically implemented as stand-alone or logically a part of some other element.

Radio Network Subsystem (RNS)

One RNC and its associated Node Bs are referred to collectively as the *Radio Network Subsystem* (RNS). The UTRAN may consist of one or more RNSs. The RNS provides all the functionality to establish, maintain, and clear radio connections. This functionality is therefore removed from the CN. In order to manage inter-RNS handovers, a defined interface exists between RNCs. This is called the l_{ur} interface.

Core Network (CN)

Many possibilities are available for the structure and form of the CN in that it may be an evolved mobile network or an evolved fixed network. Most likely, however, the CN will support both real-time and nonreal-time connections. This may be achieved through both circuit- and packet-switched technologies, or possibly packet-switched alone.

Figure 11-9 shows the proposed 3G CN architecture. One can see that the UTRAN is supporting two connections on the l_u interface linking a 3G-SGSN and an *Mobile Switching Center* (MSC).

The 3G-SGSN is connected to the *Public Land Mobile Network* (PLMN) backbone network via the G_n interface and, as with GPRS, this connection is based upon *Infrared* (IR) packet data and leaves the CN through either a 3G-GGSN or Border Gateway, which connects the Internet (G_i) and visited PLMN (G_p), respectively.

As mentioned earlier, it may be possible to do without the l_u-CS interface supporting the MSC with the introduction of *Voice over IP* (VoIP). This would enable real-time connections to run over a packet-switched technology, therefore reducing the need for traditional circuit-switched equipment. The *Third-Generation Partnership Project* (3GPP) is currently investigating

Figure 11-9
The CN in a 3G
world.

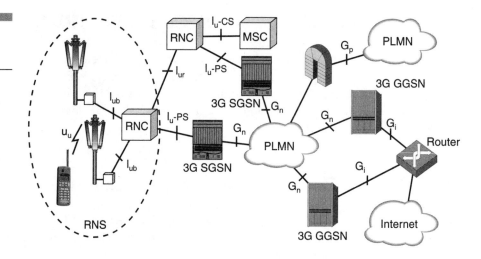

the possibility of interworking the VoIP standards (H.323) into their technical specifications.

The 3GPP is proposing the introduction of mobile IP into the CN. This
would enable a subscriber to register his or her home IP address onto any
visited network supporting mobile IP and thus negate the requirement for
multiple IP addresses. Such technology would be independent of the underlying network technology and would enable subscribers to move easily from
their LAN to a mobile environment.

Protocol Architecture

Figure 11-10 illustrates the protocol stack for the 3G CN. This protocol
stack is similar to the one being utilized by GPRS. However, it must be
noted that the protocols sitting above RLC and *GPRS Tunneling Protocol-
User Plane* (GTP-U) are for further study by 3GPP.

The main difference between the 3G protocol stack and the GPRS stack
is the lack of the *Subnetwork-Dependent Convergence Protocol* (SNDCP),
Link Layer Control (LLC), and *BSS GPRS Protocol* (BSSGP). In addition,
Frame Relay has been replaced by the *Asynchronous Transfer Mode* (ATM)
switching technology supporting the *ATM Adaptation Layer 5* (AAL5). Also,
the *Transmission Control Protocol* (TCP) will not be used when tunneling
data across the Gn interface. Instead, the *User Datagram Protocol* (UDP)
will support all connections.

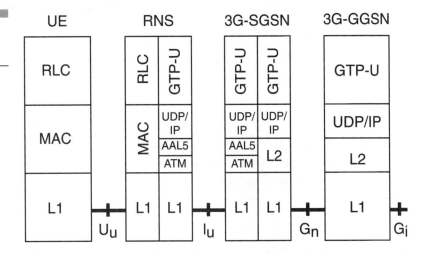

Figure 11-10
The 3G protocol stack.

UMTS

UMTS and other IMT-2000 3G mobile systems will deliver voice, graphics, video, and other broadband information directly to the user, regardless of location, network, or terminal. These personal communication services will provide terminal and service mobility on fixed and mobile networks, taking advantage of the convergence of existing and future fixed and mobile networks and the potential synergies that can be derived from such convergence. The key benefits that UMTS/IMT-2000 promises include improvements in quality and security, incorporating broadband and networked multimedia services, flexibility in service creation, and ubiquitous service portability.

Networked multimedia can be defined to include services such as

- Pay-TV
- Video- and audio-on-demand
- Interactive entertainment
- Educational and information services
- Communication services such as video-telephony and fast, large file transfer

UMTS is the European member of the IMT-2000 family of 3G cellular mobile standards. UMTS will enter the market at a time when fixed mobile integration is becoming a reality; the telecommunications, computer, and media industries have converged on IP as a shared standard; and data accounts for a significant proportion of the traffic carried by mobile networks. UMTS requirements include

- Small, low-cost packet terminals
- Worldwide roaming
- A single system for residential, office, cellular, and satellite environments
- High-speed data
  - Vehicular 144 Kbps
  - Pedestrian 384 Kbps
  - Indoor 2 Mbps

The UMTS systems, as shown in Figure 11-11, will support data rates of up to 2 Mbps and new multimedia applications over a new, wideband air interface based on CDMA techniques. Services will be supported by a wide range of terminals tailored to the requirements of voice, data, and multimedia services.

UMTS will encompass more than just cellular systems, evolving from GSM and embracing fixed networks and other wireless and wireline access technologies. Services will be globally available, delivered over the mobile, satellite, or fixed networks that provide the best accessibility for the consumer's specific location.

The current vision of most operators is that UMTS will exist as "islands of coverage" with data services supported by GPRS in areas of lower traffic

Figure 11-11
UMTS system configurations.

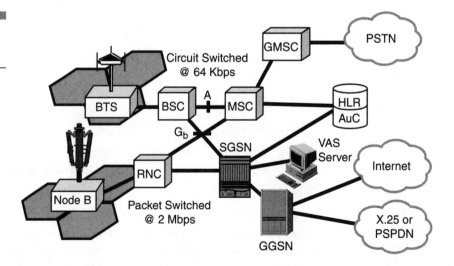

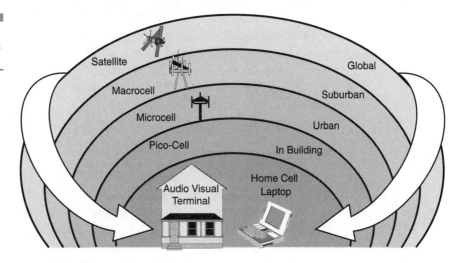

Figure 11-12
The 3G topologies as they combine.

density. Figure 11-12 shows the topologies for the configuration of the wireless future. If data demand is sufficient, it may be economical to upgrade such areas to EDGE, rather than deploy W-CDMA. Despite the apparent attractions of deploying EDGE as an incremental solution, operators will need to deploy UMTS, as only W-CDMA can support the high traffic densities encountered in the core of mature networks.

The initial release of the EDGE standard is aimed at increasing the capacity and speed of GPRS data services. The second phase of the EDGE standard will support packet voice using VoIP techniques.

User Benefits of UMTS

In line with subscribers' increasing expectations of GSM systems, UMTS will, of course, provide a very high quality of service in all environments. This will be further enhanced by the implementation of the *Adaptive Multi-Rate* (AMR) codec. In addition, users will benefit from the following features.

Seamless Global Roaming

The implementation of the Virtual Home Environment will give users the same seamless service regardless of serving network type. This means that users can access their personalized service profile through any network from any terminal, optimize the display of information, and simplify access to the key services that they use most. This programmable personality will be stored in the *Subscriber Identity Module* (SIM) card, and this will enable the same user interface to be available on any phone anywhere in the world.

High-Speed Data Services

The UMTS network will provide cost-effective data transmission with the flexibility to remain online at all times, while only paying for the amount of data received or transmitted. Terminals will always be connected to the network, e-mails could be received as soon as they are sent, and access to the intranet and Internet will be immediately available all the time with no setup delay. All this will be available at even higher data rates than those offered by GPRS systems.

Multimedia Services

New multimedia services will include video conferencing, interactive entertainment, and video transport in the case of an emergency or disaster. Multimedia technology will also make it possible to offer electronic magazines or newspapers complete with graphics and video clips.

New Innovative Applications

The involvement of new Value-Added Service Providers in the UMTS commercial model provides the opportunity for a wide range of new applications to be offered. Examples are supplementary features for traditional voice callers such as location-based services.

Telematics

Building on GPRS services, UMTS will support machine-to-machine communications in applications such as vending machine monitoring.

Increased Integration Between Fixed and Mobile Telephony Services

The increased integration of these services offers users both an increase in ease of use and increased affordability.

Increased Choice of Services

The opening up of the market for service provision and the simplification of service creation will provide users with an increased range of services from which to select. The increase in competition in the market is also expected to ensure that these services are offered to the user at an affordable price.

UMTS Future Vision

The UMTS CN will be based upon a broadband network carrying IP-based traffic. An ATM network could provide the quality of service necessary for reliable and efficient transport of multimedia data. Due to the need to support legacy interconnect options to the cell site for many years, Frame Relay remains an attractive option to maximize the efficiency of the *Base Transceiver Station* (BTS) backhaul links.

The key changes in the UMTS architecture are that

- The *Network Subsystem* (NSS) has moved to an efficient packet-based transport, using low-cost standard packet switches to route the call and signaling traffic. This also requires changes to the peripherals, such as the voice mail system, which now operates in a packet-based voice-transcoded (and thus higher voice quality) mode.
- Transcoding and data-interworking functions have moved to the periphery of the network, where it connects with other networks.

In the future UMTS network, the functions required to control the mobile network are server based and the underlying broadband network carries out the switching functions, as shown in Figure 11-13. The core platforms are built upon a common hardware and software architecture, enabling functions to be distributed as required.

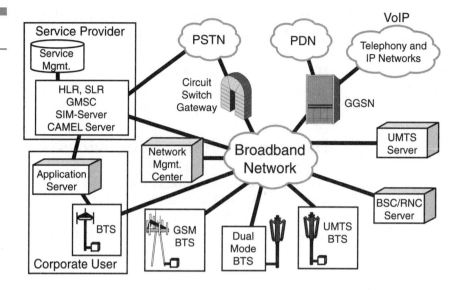

Figure 11-13
UMTS in the future.

Spectrum for UMTS

WRC-2000 identified the frequency bands 1,885 to 2,025 MHz and 2,110 to 2,200 MHz for future IMT-2000 systems, with the bands 1,980 to 2,010 MHz and 2,170 to 2,200 MHz intended for the satellite part of these future systems. CDMA is characterized by high capacity and small cell radius, employing spread-spectrum technology and a special coding scheme.

The capabilities of cdmaOne evolution have already been defined in standards. IS-95B provides *Integrated Services Data Network* (ISDN) rates up to 64 Kbps. The next phase of cdmaOne is a standard known as 1XRTT and enables 144-Kbps packet data in a mobile environment. Other features available include a twofold increase in both standby and talk time on the handset. All of these capabilities will be available in an existing cdmaOne 1.25-MHz channel. The next phase of cdmaOne evolution will incorporate the capabilities of 1XRTT, support all channel sizes (5 MHz, 10 MHz, and so on), provide circuit and packet data rates up to 2 Mbps, incorporate advanced multimedia capabilities, and include a framework for advanced 3G voice services and vocoders, including voice over packet and circuit data. Many of the steps have already been started and put in place. Table 11-2 summarizes the variations of CDMA.

Table 11-2

The Various Forms
of CDMA

CDMA Type	Description
Composite CDMA/TDMA	Wireless technology that uses both CDMA and TDMA. For large-cell licensed band and small-cell unlicensed band applications. Uses CDMA between cells and TDMA within cells.
CDMA	In addition to the original Qualcomm-invented *Narrowband CDMA* (N-CDMA) (originally just CDMA) also known in the United States as IS-95. Latest variations are B-CDMA, W-CDMA, and composite CDMA/TDMA. CDMA is characterized by high capacity and small cell radius, employing spread-spectrum technology and a special coding scheme. B-CDMA is the basis for 3G UMTS.
cdmaOne	1G N-CDMA (IS-95).
cdma2000	The new 2G CDMA *Memorandum of Understanding* (MoU) specification for inclusion in UMTS.

The cdma2000 Family of Standards

The cdma2000 family of standards includes core air interface, minimum performance, and service standards. The cdma2000 air interface standards specify a spread-spectrum radio interface that uses CDMA technology to meet the requirements for 3G wireless communication systems. In addition, the family includes a standard that specifies analog operation to support dual-mode mobile stations and base stations.

The technical requirements contained in cdma2000 form a compatibility standard for CDMA systems. They ensure that a mobile station can obtain service in a system manufactured in accordance with the cdma2000 standards. The requirements do not address the quality or reliability of that service, nor do they cover equipment performance or measurement procedures. Compatibility, as used in connection with cdma2000, is understood to mean that any cdma2000 mobile station is able to place and receive calls in cdma2000 or IS-95 systems. Conversely, any cdma2000 system is able to place and receive calls for cdma2000 and IS-95 mobile stations. In a subscriber's home system, all call placements are automatic. Similarly, call placement is automatic when a mobile station is roaming. To ensure compatibility, radio system parameters and call processing procedures are specified. The sequence of call processing steps that the mobile stations and base stations execute to establish calls is specified, along with the digital control messages and, for dual-mode systems, the analog signals that are exchanged between the two stations.

The base station is subject to different compatibility requirements than the mobile station. Radiated power levels, both desired and undesired, are fully specified for mobile stations, in order to control the RF interference that one mobile station can cause another. Base stations are fixed in location and their interference is controlled by proper layout and operation of the system in which the station operates. Detailed call processing procedures are specified for mobile stations to ensure a uniform response to all base stations. Base station procedures, which do not affect the mobile stations' operation, are left to the designers of the overall land system. This approach to writing the compatibility specification is intended to provide the land system designer with sufficient flexibility to respond to local service needs and to account for local topography and propagation conditions. cdma2000 includes provisions for future service additions and expansion of system capabilities. The release of the cdma2000 family of standards supports Spreading Rate 1 and Spreading Rate 3 operation.

The future looks rather far off when we look at the changes and the slow evolution of the standards and the spectrum allocation. However, we are on the brink of the overall changes that will escalate at a rapid pace once the momentum gets started. The use of high-speed packet-switched data communications alongside of fast voice networks will likely be ours for the taking soon. Good luck and have fun experimenting with the technologies and services.

ACRONYMS

1G	First Generation
1XRTT	A spectrum migration technology for existing IS 95 bands and systems. 1XRTT phase employs 1.25 MHz of bandwidth and is expected to provide up to twice the Radio Frequency (RF) capacity of current technologies. It is expected to support speeds of 144 Kbps for stationary and mobile applications.
IS95	Interim Standard 95
2G	Second Generation
3G	Third Generation
3XRTT	A spectrum migration technology for existing IS95 bands and systems. 3XRTT will support all channel sizes (6x, 9x, and 12x) and will use 5 MHz of bandwidth. It is expected to allow up to 2 Mbps for stationary applications
AA	Anonymous Access
AAL	ATM Adaptation Layer
AB	Access Burst
ACTS	Advanced Communication Technology Satellite
ADPCM	Adaptive Differential Pulse Code Modulation
AM or ASK	Amplitude Modulation
AMPS	Advanced Mobile Phone Service
APN	Access Point Name
ARQ	Automatic Repeat Request
ASP	Application Service Provider
ATDMA	Asynchronous Time Division Multiplexing Access
ATM	Asynchronous Transfer Mode
AuC	Authentication Center
BA	BCCH Allocation
BCS	Block Check Sequence

BEC	Backward Error Correction
BER	Bit Error Rate
BG	Border Gateway
BH	Block Header
BSC	Base Station Controller
BSN	Block Sequence Number
BSS	Base Station Subsystem
BSSAP	BSS Application Part
BSSAP+	Subset of BSSAP procedures
BSSGP	BSS GPRS Protocol (It conveys LLC PDUs between BSS and SGSN through a connectionless link [the underlying protocol is Frame Relay].)
BSSMAP	Base Station System Mobile Application Part
BTS	Base Transceiver Station
BVC	BSSGP Virtual Connection (It represents a point of interconnection between peer BSSGP entities.)
BVCI	BSSGP Virtual Connection Identifier (It is the unique identification of a BVC.)
CAGR	Compound Annual Growth Rate
CAMEL	Customized Applications for Mobile Enhanced Logic
CAP	Competitive Access Provider
CC	Control Channel
CCA	Clear Channel Assessment
CCU	Channel Codec Unit (in the MS and in the BTS)
CDG	CDMA Development Group
CDMA	Code Division Multiple Access
CDMA-W	Code Division Multiple Access-Wideband
CDPD	Cellular Digital Packet Data
CEPT	Conference of European Posts and Telecommunications
CGI	Cell Global Identity (CGI = LAI + CI)
CI	Cell Identity (It identifies one cell through one number in the network.)

CLNP	Connectionless Network Protocol
CLNS	Connectionless Network Service
CMRS	Commercial Radio Service
CO	Central Office
COCC	Constellation Operations Control Center
CONS	Connection-Oriented Network Service
CPS	Carrier Packet Solution
CRTC	Canadian Radio and Television Commission
CS	Circuit Switched
CS	Coding Scheme
CS-1	Coding Scheme 1 (data rate = 8 Kbps; 20 bytes)
CS-2	Coding Scheme 2 (data rate = 12 Kbps; 30 bytes)
CS-3	Coding Scheme 3 (data rate = 14.4 Kbps; 36 bytes)
CS-4	Coding Scheme 4 (data rate = 20 Kbps [no forward data correction]; 50 bytes)
CTIA	Cellular Telecommunications Industry Association
CU	Cell Update
DAMPS	Digital AMPS
DCA	Dynamic Channel Assignment
DCCH	Digital Control Channel
DCE	Data Circuit-Terminating Equipment
DCF	Distributed Coordination Function
DECT	Digital European Cordless Telephony
DFA	Designated Filing Area
DHCP	Dynamic Host Configuration Protocol
DIFS	Distributed Inter-Frame Spacing
DNS	Domain Name Service/Domain Name System
DS	Direct Sequence
DSAP	Destination Service Access Point
DSI	Digital Speech Interpolation technique
DSSS	Direct Sequence Spread Spectrum

DTAP	Direct Termination Application Part
DTE	Data Terminating Equipment
DWDM	Dense Wave Division Multiplexing
ECSA	Exchange Carriers Standards Association
EDGE	Enhanced Data for GSM Evolution
EDGE	Enhanced Data rates for Global Evolution
EGPRS	EDGE-based GPRS
EIA	Electronic Industries Association
EIR	Equipment Identity Register/Equipment Inventory Register
ETDMA	Extended Time Division Multiple Access
ETSI	European Telecommunications Standards Institute
FCC	Federal Communications Commission
FCS	Frame Check Sequence
FDD	Frequency Division Duplex
FDMA	Frequency Division Multiple Access
FH	Frequency Hopping
FH	Frame Header
FHMA	Frequency Hopping Multiple Access
FM or FSK	Frequency Modulation
FRAD	Frame Relay Access Device
FS	Fixed Station
FTP	File Transfer Protocol
FWRA	Fixed Wireless Radio Access
G_b	Interface between BSS and SGSN
G_c	Interface between GGSN and HLR (optional interface, if not, the GGSN can access the HLR through the SGSN)
G_d	Interface between SGSN and SMS-GMSC
G_f	Interface between SGSN and EIR
G_i	Reference point (interface between the GGSN and the external PDN—at least one for EP)
GEO	Geosynchronous orbit and one for X.25 (outside the scope of GPRS recommendations)

GGSN	Gateway GPRS Support Node (The GGSN may be considered a domain server from the Internet point of view.)
GHz	Gigahertz
GMM	GPRS Mobility Management
GMSK	Gaussian Minimum Shift Keying
G_n	Interface between GGSN and SGSN
G_p	PLMN to PLMN (that is, roaming)
GPDS	General Packet Data Service
GPRS	General Packet Radio Service
GPRS	General Packet Radio System
GPS	Global Positioning System
G_r	Interface between SGSN and HLR
GR	GPRS Register (conceptually art of HLR)
G_s	Interface between SGSN and MSC (optional interface)
GSM	Global Services for Mobile
GSN	GPRS Support Node
GTP	GPRS Tunneling Protocol (It enables PDU to be tunneled between the GGSN and the SGSN through an IP network.)
HDML	Hand-Held Device Markup Language
HF	High Frequency
HLR	Home Location Register
HSCSD	High-Speed Circuit-Switched Data
Hz	Hertz
IETF	Internet Engineering Task Force
IIF	Interworking and Interoperability Function
IMEI	International Mobile Equipment Identity
IMSI	International Mobile Subscriber Identity
IMT-2000	International Mobile Telecommunications 2000
IMTS	Improved Mobile Telephone Service
IN	Intelligent Networking
IP	Internet Protocol (network layer protocol providing a connectionless service)

IPv4	IP version 4
IPv6	IP version 6
IPR	Intellectual Property Rights
IR	Infrared
IrDa	Infrared port
IS	Information Service
ISDN	Integrated Services Data Network
ISM	Industrial Scientific Medical
ISP	Internet Service Provider
ITU-T	International Telecommunication Union-Telecommunication Standardization Sector
IWF	Interworking Function
IWU	Interworking Unit
kHz	Kilohertz
LA	Location Area
LAN	Local Area Network
LAC	Location Area Code
LAI	Location Area Identity (LAI = MCC + MNC + LAC)
LAPD	Link Access Procedure for the D channel
LEO	Low Earth Orbit
LLC	Logical Link Control (It provides an OSI Layer 2 logical connection between the mobile station and the SGSN; it is an LAPD like L2—sequential order of delivery and detection and recovery of errors and flow control.)
LOS	Signal in a Line-Of-Sight
LPC	Linear Predictive Coding
MAC	Medium Access Control
MAHO	Mobile-Assisted Handoff
MAP	Mobile Application Protocol (users protocol based on SS7, which manages communications between NSS equipment)
MCC	Mobile Country Code
MEO	Mid-Earth Orbit

MF-TDMA	Multi-Frequency Time Division Multiple Access
MGCP	Media Gateway Control Protocol
MHz	Megahertz
MIS	Management Information Service (or System Support)
MM	Mobility Management
MNC	Mobile Network Code
MoU	Memorandum of Understanding
MS	Mobile Station (It is made of TE and MT, linked by the R reference point.)
MSA	Metropolitan Service Area
MSC	Mobile Switching Center
MTP	Mobile/Message Transfer Part
MT	Mobile Termination/Mobile Terminal
MTSO	Mobile Telephone Switching Office
NADC	North American Dual-Mode Cellular
N-AMPS	Narrowband Advanced Mobile Phone Service
NAM	Number Assignment Module
NB	Normal Burst
NCH	Notification Channel
NIF	Network Interface Facility
NMS	Network Management System
NOC	Network Operations Center
NOCC	Network Operations Control Center
N-PDU	Network Protocol Data Unit
NSAPI	Network layer Service Access Point Identifier (In the MS, it identifies the PDP-SAP; in the SGSN and GGSN, it identifies the PDP context associated with a PDP address.)
NSS	Network Subsystem
NS VCI	Network Service Virtual Connection Identifier
OMC	Operations Maintenance and Control
OMC-D	Operations Maintenance and Control-Data

OMC-R	Operations Maintenance and Control-Radio
OSS	Operation and Support System
PACCH	Packet Associated Control Channel
PAGCH	Packet Access Grant Channel (downlink only, used to allocate one or several PDTCHs—part of PCCCH)
PAN	Personal Area Network
PBCCH	Packet Broadcast Control Channel
PBS	Personal Base Station
PC	Power Control
PCCCH	Packet Common Control Channel (includes PPCH, PRACH, PAGCH, and PNCH)
PCIA	Personal Communications Industry Association
PCM	Pulse Coded Modulation
PCMCIA	PC Modular Communication Interface Adapter
PCS	Personal Communications Service
PCU	Packet Control Unit (In BSS—BTS or TCU—or SGSN, it acts as an LLC relay between the mobile station and SGSN. The main function is to manage channel and radio link control.)
PCUSN	Packet Control Unit Support Node
PDA	Personal Digital Assistant
PDC	Personal Digital Cellular
PDCH	Packet Data Channel (physical channel dedicated to packet logical channels only)
PDH	Plesiochronous Digital Hierarchy
PDN	Packet Data Network
PDP	Packet Data Protocol (For example, IP or X.25 PDP address —it can be IPv4, Ipv6 or X.121 address.)
PDTCH	Packet Data Traffic Channel (All packet data traffic channels are unidirectional, either uplink (PDTCH/U) for a mobile-originated packet transfer, or downlink (PDTCH/D) for a mobile-terminated packet transfer.)
PDU	Packet Data Unit/Protocol Data Unit
PHY	Physical

PL	Physical Link
PLCP	Physical Layer Convergence Protocol
PLMN	Public Land Mobile Network
PM or PSK	Phase Modulation
PMD	Physical Medium Dependent
PMR	Private Mobile Radio
PN	Pseudo-random Noise
PNCH	Packet Notification Channel (downlink only, used to notify the mobile station of PTM-M call—part of PCCCH)
POP	Point of Presence
POTS	Plain Old Telephone Service
PPCH	Packet Paging Channel (downlink only, used to page the mobile station—part of PCCCH)
PPF	Paging Proceed Flag (In the SGSN, it is cleared when the mobile reachable timer expires. MM and PDP contexts are kept, but no more paging.)
PRACH	Packet Random Access Channel (uplink only, used to request allocation of one or several PDTCHs for uplink or downlink direction—part of PCCCH)
PSK	Phase Shift Keying
PSTN	Public Switched Telephone Network
PTCCH/D	Packet Timing advance Control Channel Downlink (It's used to transmit timing advance updates for several mobile stations—up to 16. One PTCCH/D is paired with several PTCCH/Us.)
PTCCH/U	Packet Timing advance Control Channel Uplink (It's used to transmit random access bursts to enable estimation of the timing advance for one mobile station in packet transfer mode.)
PTM	Point To Multipoint
PTM-G	Point To Multipoint-Group (call to a dedicated group of users)
PTM-M	Point To Multipoint-Multicast (in a dedicated area, not specified in GPRS phase 1)
P-TMSI	Packet-TMSI

PTO	Public Telephony Operator
PTP	Point To Point
PTP-CLNS	PTP-Connectionless Network Service
PTP-CONS	PTP-Connection Oriented Network Service
PVC	Permanent Virtual Circuit
PVG	Passport Voice Gateway
QAM	Quadrature with Amplitude Modulation
QoS	Quality of Service
QPSK	Quadrature Phase Shift Keying
R	Reference point (At this access point non-ISDN compatible bearer services may be accessed, for example, ITU-T, X-, and V-series recommendation.)
RA	Routing Area (subset of one, and only one, LA—served by only one SGSN). A group of one or many cells creates a routing area.
RAC	Routing Area Code
RAI	Routing Area Identity (RAI = MCC + MNC + LAC + RAC)
RAN	Radio Access Network
RF	Radio Frequency
RITL	Radio In The Loop concept
RLC	Radio Link Control
RLP	Radio Link Protocol
RNC	Radio Network Controller
RTT	Radio Transmission Technology
SACCH	Slow Associated Control Channel
SAP	Service Access Point
SAPI	Service Access Point Identifier
SCCP	Signal Connection Control Part
SDH	Synchronous Digital Hierarchy (In Europe, corresponding to SONET in North America, the standard frame is called STM-1: Synchronous Transfer Module level 1.)
SDU	Service Data Unit
SGSN	Serving GPRS Support Node (It performs MM, authentication procedures, and routes packet data.)

SIG	Special Interest Group
SIG	SS7-to-IP Gateway
SIM	Subscriber Identity Module
SM	Session Management
SMR	Specialized Mobile Radio
SMS	Short Message Service
SMS-SC	Short Message Service-Service Center
SMS-GMSC	Short Message Service-Gateway MSC
SMTP	Simple Mail Transfer Protocol (e-mail protocol)
SND	Sequence Number Downlink
SNDCP	Subnetwork-Dependent Convergence Protocol (It maps the network protocols to best fit the underlying GPRS transmission capabilities and covers ciphering, segmentation, and compression.)
SN-PDU	SNDCP-Protocol Data Unit
SNR	Signal-to-Noise Ratio
SNU	Sequence Number Uplink
SONET	Synchronous Optical Network
SSAP	Source Service Access Point
SVC	Switched Virtual Circuit
TA	Timing Advance
TA	Terminal Adaptor
TACS	Total Access Communications Service
TACS	Total Access Control System
TAF	Terminal Adaptation Function
TAI	Timing Advance Index
TBF	Temporary Block Flow
TBFI	Temporary Block Flow Indicator
TCAP	Transaction Capabilities Application Part
TCH	Traffic Channel
TCP	Transmission Control Protocol
TDD	Time Division Duplex
TDM	Time Division Multiplexing

TDMA	Time Division Multiple Access
TE	Terminal Equipment
TFI	Temporary Flow Identity (for RLC transfer purpose)
TFI	Temporary Frame Indicator
TI	Transaction Identifier
TIA	Telecommunications Industry Association
TID	Tunnel Identifier (Used by GTPs between GSNs to identify a PDP context; it consists of an IMSI and an NSAPI.)
TLLI	Temporary Logical Link Identity (It identifies the logical link between the mobile station and SGSN—derived from P_TMSI.)
TMSI	Temporary Mobile Subscriber Identity
TRAU	Transcoder and Rate Adapter Unit
TSC	Trunking System Controller
TSC	Transit Switching Center
TTML	Tagged Text Markup Language
UDP	User Data Protocol (User Datagram Protocol is located in the transport layer, but in a connectionless mode and with practically no functionality.)
UHF	Ultra High Frequency
UI	Unnumbered Information
UMTS	Universal Mobile Telephone System
UP	Unified Protocol
USF	Uplink State Flag (It is used on PDCH to allow multiplexing of radio blocks from a number of mobile stations. USF is used in dynamic and extended dynamic medium access modes. USF is used only in downlink direction; it is made of 3 bits.)
USSD	Unstructured Supplementary Services Data
UTRA	Universal Terrestrial Radio Access
UTRAN	Universal Terrestrial Radio Access Network
UWCC	Universal Wireless Communications Consortium
VHF	Very High Frequency
VLR	Visiting Location Register

VoIP	Voice-over Internet Protocol
VPN	Virtual Private Network
VSAT	Very Small Aperture Terminal
VSELP	Vector Sum Excited Linear Predictor
WAN	Wide Area Network
WAP	Wireless Application Protocol
WARC	World Administrative Radio Conference
WCAP	Wireless Competitive Access Provider
W-CDMA	Wideband-CDMA
WCS	Wireless Communications System
WI	Wireless Internet
WIT	Wireless Intelligent Terminal
WLAN	Wireless LAN
WLL	Wireless Local Loop
WWAN	Wireless Wide Area Network

INDEX

Does Your Organization Need On-site, Instructor-led Training?
1-800-322-2202—www.tcic.com

TCIC specializes in Technology Training on the following topics (these are just some of the titles):

- Data Communications
- Voice Communications
- LAN/WAN Networking
- Wireless Communications
- Introduction to GSM

- GPRS Overview
- Understanding ATM
- Frame Relay
- Signaling System 7 (SS7)
- Voice over IP

Don't have the time to spend days away from your job? We now offer some of the best Computer-Based Training (Multimedia) in the industry:

- Voice Communications Demystified
- Data Communications Demystified
- LAN/WAN Networking
  - Module I
  - Module II
  - Module III

- T1 Networking
  - Module I
  - Module II
- xDSL
- SS7

Available soon on CD: SONET and SDH, SS7, ATM, Cable Modems and Frame Relay. (These titles are also available for your Intranet in html format.)

Keynote Speeches:

Hire the author for your next keynote! Bates is knowledgeable, at the same time animated, motivational, and engaging! He has ignited and charged audiences around the globe!

Call now **1-800-322-2202** or visit our website **www.tcic.com**

WWW.TCIC.COM

1-800-322-2202
info@tcic.com